Intelligent Visual Inspection

Intelligent Engineering Systems Series

Series Editor: Cihan H. Dagli
Department of Engineering Management
University of Missouri-Rolla, USA.

Engineering Systems of the next century need to be autonomous to meet the challenge of flexibility and customized design requirements imposed on manufacturing and service systems by the global economy. There is a need to build 'intelligent' components for engineering systems currently available today. The term 'intelligent' in this context indicates physical systems that can interact with their environment and adapt to changes both in space and time. This is achieved by their ability to manipulate the environment through self-awareness and perceived models of the world, based on both quantitative and qualitative information. The emerging technologies of artificial neural networks, fuzzy logic, evolutionary programming, chaos, wavelets, fractals, complex systems, and virtual reality provide essential tools for designing such systems.

This series is established to disseminate recent developments in this area to researchers and practicing engineers. The books in the series cover intelligent engineering architectures that integrate and/or enhance the current and future technologies necessary for developing intelligent engineering systems, while illustrating the real life application of these architectures. The intelligent engineering systems design and operation cut across a diversity of disciplines, namely: Manufacturing, Electrical, Computer, Mechanical, Bio-Medical, Civil Engineering and other related fields such as Applied Mathematics, Cognitive Sciences, Biology and Medicine. The series will feature books on a number of topics including:

- *Intelligent Engineering Architectures*
- *Neural Networks*
- *Fuzzy Systems*
- *Evolutionary Programming*
- *Automated Inspection*
- *Mechatronics*
- *Wavelets, Fractals, Chaos*
- *Complex Systems*
- *Virtual Reality*
- *Process Monitoring and Control*
- *Scheduling*
- *Automated Design*

Some of the titles are more theoretical in nature, while others emphasize real life applications. Books dealing with the most recent developments will be edited by the leaders in their particular fields. Recognized author's contributions to the series will be welcomed in more established areas.

We are confident that the series will be appreciated by researchers, graduate students in engineering schools and research laboratories, practicing engineers and scientists in other related fields.

Intelligent Visual Inspection

Using artificial neural networks

Ryan G. Rosandich

Department of Industrial Engineering
University of Minnesota-Duluth
10 University Drive
Duluth
MN 55812
USA

CHAPMAN & HALL

London · Weinheim · New York · Tokyo · Melbourne · Madras

Published by Chapman & Hall, 2–6 Boundary Row, London SE1 8HN, UK

Chapman & Hall, 2–6 Boundary Row, London SE1 8HN, UK

Chapman & Hall GmbH, Pappelallee 3, 69469 Weinheim, Germany

Chapman & Hall USA, Fourth Floor, 115 Fifth Avenue, New York, NY 10003, USA

Chapman & Hall Japan, ITP-Japan, Kyowa Building, 3F, 2-2-1 Hirakawacho, Chiyoda-ku, Tokyo 102, Japan

Chapman & Hall Australia, 102 Dodds Street, South Melbourne, Victoria 3205, Australia

Chapman & Hall India, R. Seshadri, 32 Second Main Road, CIT East, Madras 600 035, India

First edition 1997

Softcover reprint of the hardcover 1st edition 1997

Typeset in 10/12 pt Palatino by Cambrian Typesetters, Frimley, Surrey

ISBN-13: 978-1-4612-8510-6 e-ISBN-13: 978-1-4613-1201-7
DOI: 10.1007/978-1-4613-1201-7

A catalogue record for this book is available from the British Library

Library of Congress Catalog Card Number: 96–71259

Printed on permanent acid-free text paper, manufactured in accordance with ANSI/NISO Z39.48-1992 and ANSI/NISO Z39.48-1984 (Permanence of Paper).

Contents

Preface

Over the past several years a great deal of research has been carried out in the areas of artificial vision and neural networks. Although much of this research has been theoretical in nature, many of the techniques developed through these efforts are now mature enough to be used in practical applications.

Concurrently, the opening of worldwide markets to many products has forced manufacturing companies to compete on a global basis in recent years. This high level of competition between manufacturers has led to rapid developments in the areas of computer-integrated manufacturing, flexible manufacturing, agile manufacturing and intelligent manufacturing. These developments have in turn generated a need for intelligent sensing and decision making systems capable of automatically performing many tasks traditionally done by human beings. Visual inspection is one such task, and there is a need for effective automated visual inspection systems in today's competitive manufacturing environment.

In this book, the author seeks to accomplish two things. First, to collect and organize information in the artificial vision and neural network fields and to present that information in an informative, clear manner without excess mathematical detail. Second, to help bridge the gap between the theoretical research that has been done and the present needs of manufacturing companies by demonstrating how several recently developed techniques can be implemented via existing hardware to solve practical automated inspection problems. To this end the book is organized into four sections. Section One introduces and defines the problem of intelligent visual inspection in manufacturing environments. Section Two reviews fundamental research in the areas of artificial vision and neural networks. Section Three covers in detail the practical design of artificial vision systems based on neural networks, and Section Four presents case studies which show how these systems can be applied to current and future real-world inspection problems.

This book is intended to appeal to electrical, mechanical, industrial and manufacturing engineers, computer scientists, technicians, and managers employed by manufacturing companies that are interested in the potential of intelligent visual inspection systems. It should also appeal to manufacturers of machine vision systems, and to consulting engineers and system integrators engaged in the design and installation of such systems.

The book is intended for academic appeal as well. It should serve as a reference book for researchers and graduate students interested in the areas of artificial vision or neural networks, and it should also be a valuable reference book for courses with content related to these topics. Such courses could be part of a curriculum in computer science or electrical, mechanical, manufacturing, or industrial engineering. Finally, management programs may also be interested in the book as a reference for graduate courses in quality control.

I would like to thank those who have helped make this book become a reality. First I would like to thank my former department at the University of Kansas for tolerating my continued work on this manuscript in the presence of a busy teaching and research schedule. They also provided, at various times, equipment and facilities to support some of the research described in this book. I would also like to thank Dr. Cihan H. Dagli of the Department of Engineering Management at the University of Missouri-Rolla for providing leadership and advice during my Ph.D. program which ultimately led to much of the research described in this book, and for making the contacts necessary to get this manuscript published. In addition, I would like to thank the University of Missouri-Rolla for providing financial support via the Chancellor's Fellowship for my pursuit of the Ph.D. degree. Finally, I would like to thank Cosmin Radu, a graduate student at the University of Missouri-Rolla, for conducting research and providing information pertinent to Chapters 7 and 11, and Herb Tuttle of the University of Kansas for providing information for Chapter 12.

Part One

Introduction

1

Intelligent manufacturing

1.1 INTRODUCTION

You may not realize it, but you are performing a remarkable feat by reading this page. You are visually analyzing and interpreting a series of small black shapes on a white page, associating those shapes with words that you know, and chaining those words into meaningful thoughts. Along with assimilating information, you are also performing an inspection of sorts, constantly judging the text in front of you. You would probably notice a spelling mistrake, or if the sizes of the letters changed, or if word spacing became erratic. You also notice and judge more subtle things like grammar, the flow of the narrative, and the quality of the information presented.

The point is that human beings are very visually oriented and naturally critical. They also have an uncanny ability to notice small variations in appearance. If you hold this page at arm's length, your attention will probably be drawn to the few characters or word spacings in the previous paragraph that are not sized consistently. Out of the hundreds of letters and words on this page, you notice the few odd ones at a glance.

This perceptive ability makes humans difficult customers to produce products for, but they are the only customers that are available! Fortunately, these same qualities also make human beings good inspectors of products. Because of their level of intelligence, however, humans quickly become bored with repetitive tasks like inspection. They also fatigue rapidly when tasks require constant concentration and little physical activity, and their performance deteriorates.

The obvious answer is an automated visual inspection system that is as perceptive and selective as a human being, but which can operate without boredom or fatigue for days on end. The second part of this specification is easier to meet than the first. Automated systems of many varieties have been constructed to perform tasks formerly conducted by humans, including visual inspection. The problem with automated inspection systems is revealed when we measure them against the first part of the above specification, the part that says the system should be as perceptive and selective as a human operator. Current systems are typically designed for a specific task, and lack the intelligence and flexibility of human inspectors. This intelligence and flexibility must be built into automated visual inspection systems if they are to approach the level of performance routinely achieved by human inspectors.

1.2 DEFINITION OF INTELLIGENCE

There have been many efforts to define intelligence. The desire to determine what intelligence is, how it originates, and what fundamental processes underlie it is evident in activities such as the study of intelligent creatures, the modeling of biological intelligence, and the attempted creation of intelligent machines. A recent attempt has been made to tie all of this research together into a general theoretical model of intelligence (Albus, 1991). In this theoretical model, intelligence is defined as the ability of a system to sense its environment, make decisions, and control action. This definition implies that some form of knowledge representation is used to model the state of the environment, and that reasoning is conducted on that knowledge in order to plan future actions.

In its simplest form, intelligence is control. Decisions are made based on knowledge of the world and actions are taken (open-loop control), or actions are taken and, based on the results of those actions, decisions are made and new actions are taken (closed-loop control). Intelligence, however, is more complicated than that. Albus explains that an intelligent system has the ability to act appropriately in an uncertain environment, where appropriate behavior is that which increases the probability of the intelligent system achieving its goals. Goals are not simple but rather are arranged in a hierarchy, with higher levels representing longer-term goals. Short-term or immediate goals support long-term goals, in that achieving a short-term goal improves the probability of eventually achieving long-term goals. Many people have postulated that biological creatures have the ultimate long-term goal of propagating their species, in other words of reproducing and passing along their genes to the next generation.

Intelligence is often placed on a scale as well, measuring the intelligence of one creature or system against another. One criterion for such a comparison is the time span over which the creature or system plans its actions. Creatures or systems that exhibit organized long-range planning are generally considered more intelligent than those that act only on short-term goals. An interesting comparison in this regard is the well-known long-term business strategy employed by the Japanese versus the American style of emphasizing short-term financial results. It is left to the reader to determine which of these behaviors appears more intelligent.

Learning is another aspect of intelligence. Although learning is not a prerequisite for intelligent behavior, systems that employ learning have many advantages over those that do not. Learning systems have the ability to increase their level of intelligence over time, and to adapt to changing environments. Without learning, beings exposed to a changing environment would naturally appear to become less intelligent with time, as their fixed knowledge base becomes less and less appropriate to their situation. In the biological world beings that rely on original knowledge are said to act on instinct, whereas learning creatures modify their behavior based on their experiences. The latter are almost always more successful in the long run. Natural selection, too, provides a sort of long-term learning. Because creatures that behave in an advantageous manner are more likely to survive and reproduce, the species as a whole slowly 'learns' to achieve greater success as generations pass.

Intelligence is a group phenomenon as well as an individual trait. It is often advantageous for groups of systems or beings to act in a coordinated manner, pursuing a common goal rather than individual goals. This type of cooperation requires communication and global planning, two attributes that are usually associated with higher intelligence. Albus goes even further with this idea in stating that communication arose in intelligent individuals only because of this need to cooperate, and furthermore that the level of communication that develops between creatures directly reflects the complexity of the messages that they must share in order to facilitate this advantageous cooperation. An analogy can be drawn here with manufacturing systems, in which the desire for truly flexible and intelligent factories has instigated increasing levels of integration.

Intelligent behavior involves a system's ability to assess its present situation and then, based on that assessment and on accumulated knowledge, to plan and execute behavior that will increase the probability of the system reaching its goals. The following building blocks are required to build an intelligent system (Albus 1991):

1. *Sensors* Sensors receive the inputs to an intelligent system, which are usually measurements of physical quantities. Visual sensors may measure brightness or color, tactile sensors may measure force, torque, or position, dynamic sensors may measure velocity, acceleration or vibration, and analytical sensors may detect things like smell, taste, or temperature. Sensors can also be turned inward on the intelligent system itself to monitor its internal condition.

2. *Sensory processing* Sensory observations are compared to expectations by the sensory processing system. Expectations are generated using the internal world model as a reference. When observations confirm expectations the meaning of those observations is known, and when unexpected observations occur unique objects or events are identified and learning or adaptation occurs. Sensory processing involves processes like understanding speech and interpreting visual scenes.
3. *World modeling* The world model is an internal estimate of the current state of the external world. The world model contains data about the world, and a simulation capability that generates expectations from current states. The world model is updated whenever novel experiences are detected by the sensory processing system, and it provides input to the value judgement and behavior generation systems so they can make intelligent plans and choices. The world model represents the past, present, and likely future behavior of the external environment.
4. *Value judgement* Value judgements are required in order to determine what is good or bad for the intelligent system. Evaluating both the current state of the world and predictions for the future, the value judgement module computes costs, risks, and benefits of current and probable future situations. Values are consulted to judge the current state of the environment and to evaluate various possible plans of action.
5. *Behavior generation* The behavior-generating system selects goals, makes plans, and executes tasks. High level plans are hierarchically decomposed into subplans and finally individual tasks that can be performed by the system. Plans are communicated to the world model which predicts their effects, and the value judgement system evaluates the results. Plans with the greatest likelihood of progressing toward the goals of the intelligent system are then executed.
6. *Actuators* The actuators carry out the plans of the behavior-generating system. Tasks are communicated to the actuators, which physically alter the external environment in some way. Actuators enable exploration and movement around the environment, as well as the ability to modify the environment through the application of force or energy.

The intelligent vision systems that are described in this book can comprise a significant portion of an intelligent system. The cameras, image processing techniques, object representation and learning methods described in the following chapters serve well as the sensors, sensory processing, and world modeling segments of intelligent systems. These segments allow an intelligent system to establish a correlation between objects or events in the real world and representations in the internal world model. The introduction of learning to these segments also enables the system to update the world model based on sensory input.

1.3 ROLE OF VISION IN INTELLIGENCE

The desire to build intelligent artificial systems has led researchers to model intelligent biological systems. The most obvious of these attempts is the science of artificial neural networks, a discipline involved with imitating intelligent biological structures at the cellular level in an effort to reproduce some of their impressive capabilities. Biological intelligence and information processing are vastly different to traditional computer processing, with a huge number of massively interconnected but rather simple processors (neuron cells) being favored in biological systems over the current man-made approach which favors a small number of powerful sequential processors. The development of artificial neural networks seeks to replace the predominately sequential approach to information processing in favor of more biologically faithful techniques. Although this field is still rather new, many advances and successes have been recorded, particularly in the areas of optimization and pattern recognition. Artificial neural systems hold the promise of generating truly intelligent behavior without long processing times. Many artificial neural network techniques have recently been making the transition from theoretical development to practical application (Dagli *et al.*, 1994; Dagli *et al.*, 1995), and the work described in this book represents an attempt to further that effort.

The seemingly 'natural' ability of artificial neural systems to perform pattern recognition tasks has made them popular in that domain. The development of artificial vision systems based on neural networks, an effort that has been seriously pursued for over 10 years, is one example of this popularity. The phrase 'artificial vision systems' as used here encompasses all man-made vision systems, and includes the categories of research known as computer vision, machine vision, robot vision, pattern recognition, image processing and automated visual inspection. The study of artificial vision systems falls into the larger class of research known as artificial intelligence, which could in turn be included as a member of an emerging study area called artificial life.

Vision is thought by many to be a human being's most important sense. More processing area in the brain is dedicated to visual information processing than to any other sense, and it is possible that the high state of development of the human visual system is in part responsible for the dominance of the human species on Earth. In order for the development of intelligent machines to be a success, the development of effective artificial vision systems must be a high priority. Machines endowed with effective vision systems would have almost an unlimited number of uses, particularly in manufacturing environments. It is also possible that they could shed some additional light on the nature of intelligence itself.

1.4 MODERN MANUFACTURING SYSTEMS

1.4.1 Cost-based manufacturing

The economic prosperity in the United States that followed World War II created a tremendous demand for manufactured products. Since the primary source of competition between manufacturers was selling price, it became mandatory to reduce manufacturing costs in order to generate higher profits. This pressure for cost reduction led to a great deal of standardization and automation in manufacturing, efforts that focused on increasing production rates, reducing labor used and reducing costs. The consumer benefitted from the availability of previously unheard-of quantities of low-cost products.

All was not well in American manufacturing companies, however. The huge demand for products led to a 'sellers market' in which companies considered the demand for their products to be practically infinite. Companies believed that any reduction in cost or increase in production translated into higher profits. The internal focus created by the continuing efforts at cost reduction caused companies to ignore other factors like product quality and customer needs. Product quality suffered at many companies, as products were produced on unattended high-speed machines and the parts and materials used were the cheapest available. Engineering efforts were more likely to be focused on cost reduction than on the development of new products or the improvement of current products.

1.4.2 Quality-based manufacturing

In the 1970s the market for manufactured goods started to become global. American manufacturing companies found themselves in competition with companies from Japan and elsewhere, and the variety of products available in the marketplace increased yet again. Consumers in many countries had a choice of an unprecedented variety of foreign and domestic products, and this choice led to a shift in market conditions. Companies that previously considered the market to be infinite found themselves in a competitive market with increasingly choosy customers.

This market shift put another pressure on manufacturers. In addition to holding costs down, they had to be able to supply a product of high enough quality to compete in the market. This pressure led companies to increase the emphasis on quality, and virtually every manufacturing company instituted some kind of quality program. Emphasis was placed on statistical quality control for monitoring the production process. Quality teams, quality circles, and quality managers became commonplace, and phrases such as 'total quality management' and 'continuous improvement' were used to describe company philosophies. Japanese companies embraced quality principles early on, and they were commonly believed to be ahead in this regard. Companies from the United States and other countries were not far behind, however, and the level of quality available in the marketplace increased dramatically.

1.4.3 Time-based manufacturing

By the beginning of the 1990s many manufacturing companies were capable of providing high-quality, low-cost products to the global marketplace. Some companies were still doing significantly better than others, however. Consumers had high opinions of these companies, and their products enjoyed consistently higher sales. Many of these companies had one thing in common, something best described as **responsiveness**. They responded quickly to customer needs and desires by rapidly introducing new and different products. They responded to shifting demand by quickly adjusting production. They responded quickly to problems and changing conditions in their factories by teaming with equipment suppliers, material suppliers, labor and customers. It would seem that a new kind of pressure had been applied to manufacturing firms – time pressure.

Consumers were not satisfied with products that were just high in quality and low in cost. They wanted products to meet their needs exactly, and to be available quickly. Consumers wanted to order a product to their specifications and have it delivered with the same speed as standard products. They demanded an unprecedented diversity in product lines, a diversity which pushed quantities of identical products so low as to preclude automated mass production. Instead of offering 'any color as long as it is black' companies now had to offer any color.

'Agile manufacturing' became a popular goal. The entire manufacturing organization had to be agile enough to respond to constantly changing conditions and demands. Business processes were re-engineered to eliminate inefficiency and waste, and companies benchmarked themselves against the best in the business. The concept of 'flexible manufacturing' was promoted, a concept which theoretically allowed companies to produce a variety of products at high rates on a single production line. Many of these concepts are still evolving, and it remains to be seen what a successful manufacturing company of the future will look like. One thing seems certain, however. High quality, low cost and responsiveness are unlikely to keep the market satisfied forever.

1.4.4 The future: information-based manufacturing?

It has been said that we are now entering an information age, an age where information is more important than physical goods. Knowledge is power, and those with the most information at their disposal will be the most powerful. The meteoric rise of information technology companies such as Microsoft and Netscape would seem to support this hypothesis.

What effect will this emphasis on information have on manufacturing? Perhaps the next pressure to be applied to manufacturing companies will be pressure for information. Customers will want information about products, i.e. data such as dates of manufacture, options included, operating instructions and sources of replacement parts. The simple printed owner's manual that currently ships with most products will no longer be adequate, and it will be replaced by a sophisticated, multimedia information package. Customers will demand videotapes that show how to install or operate a product, or interactive CD-ROMs for training on product use. Companies will have to supply telephone, facsimile, and e-mail links to give customers access to product and technical information. Internet web pages will provide customers with the latest product information, as well as information on recalls, compatibility issues, and product changes or upgrades.

Companies that supply raw material and parts to manufacturing companies are not exempt from the information revolution either. To become more effective, manufacturing companies must ask many questions of their suppliers. Which parts vary from the standard, and in what way? Where are defects in raw materials located, so that they can be avoided most effectively? What is the exact physical and chemical composition of a raw material, so that a production process can be adjusted optimally? The answers to all of these questions will come from detailed information about parts and raw materials.

There are already many examples of the integration of information into products. Computer hardware and software companies are judged in part by the responsiveness of their technical support services, typically a 24-hour help line. Copiers diagnose themselves, and often have an option to phone for service automatically. It is not unusual for products to ship with computer disks or videotapes containing information on the product and its use.

Sensitive military products and products for the nuclear industry provide excellent insight into what the future may bring. The level of documentation required in the production of military goods is well known, and is perhaps representative of the future of manufacturing. In the nuclear industry, which has similar requirements, it is not unusual for the production of product documentation to cost more than the production of the product itself. For safety reasons, a paper (or electronic) trail is kept, documenting the time, date and place of manufacture, the sources of all raw materials, and detailed test data on those materials, to mention just a few requirements. Manufacturing companies of the future may have to become very adept at managing large amounts of information if this situation becomes more commonplace in the broader market.

Companies will also have to take advantage of developments in information technology to streamline their production processes. The term **virtual manufacturing** is used today to describe manufacturing systems that are actually combinations of systems owned and operated by several different companies. The various participating companies may represent engineers and designers, raw material suppliers, manufacturers of various parts, assemblers, packagers, and financial and accounting people. Virtual manufacturing systems will be created to produce a product or products and will then be modified or destroyed. The participants in the virtual manufacturing enterprise may be geographically widely distributed and may represent a variety of traditional industries. Information links will be created between the participants so that production, quality, and accounting data can be exchanged rapidly, and so that the manufacturing process can be coordinated.

1.5 CONSIDERATIONS IN INTELLIGENT MANUFACTURING SYSTEMS DESIGN

So far, this chapter has presented the topics of intelligence and manufacturing separately, but the two must be combined in order to enable manufacturing companies to compete in the future marketplace. Manufacturing systems need to become intelligent systems that make plans and pursue goals, rather than simple automatons that punch out products. Advanced sensors must be in place so that manufacturing systems can detect, analyze, and react to changes in the environment. It is no longer adequate simply to discard products not manufactured properly. Systems must be self-correcting, reacting to production problems by adapting intelligently with little human intervention. Systems must also be able to adapt to the production of a variety of products automatically, without manual changes to hard tooling. In short, the desirable attributes of intelligence and flexibility must be built into manufacturing systems.

Intelligent and flexible manufacturing systems will have a greater reliance on advanced sensors and intelligent subsystems than the manufacturing systems of today. In the chapters that follow, the theoretical development and practical application of intelligent artificial vision systems are discussed. The systems covered in this book are intended primarily to perform tasks now predominately done by humans, tasks like visual inspection and workpiece identification. Only when these intelligent visual tasks can be done by machines will automated manufacturing systems begin to exhibit the flexibility and adaptability routinely demonstrated by manually operated systems.

2

Intelligent visual inspection

2.1 VISION SYSTEMS FOR INTELLIGENT MANUFACTURING

Advanced sensors are an essential part of modern intelligent manufacturing systems. Manufacturing paradigms such as flexible manufacturing allow firms to manufacture a wide variety of products on a single system, in any product mix. They also enable the rapid introduction of new products, quick changeover between different products, and the ability to manufacture products in any random sequence. Artificial vision systems can help companies meet these flexibility requirements (Hollingum, 1984).

Machine-based vision systems can identify materials and components, and locate and orient parts prior to assembly. They can also assist with intelligent manufacturing by precisely locating mating points or machinery load points, or by verifying proper assembly. Finally, machine vision can perform inspection and measurement tasks during manufacture and prior to shipment. It is hard to imagine the automation, without the aid of machine vision, of many tasks now performed by humans in manufacturing plants.

The first crude machine vision systems appeared in industry in the 1950s with the advent of television, and they have become increasingly more capable and complex as video and computer technology have improved (Zuech, 1988). Early systems were primarily simple inspection and measurement devices, whereas more recent systems perform tasks such as object identification and intelligent robot guidance.

Applications of artificial systems in manufacturing fall into several broad categories (Zuech and Miller, 1989). One rather mature application category is optical character recognition. The need to sort millions of pieces of mail daily has driven the development of these systems to a high level, but they are still far from perfect. The techniques developed for mail sorting have also been applied to manufacturing, in systems that do everything from reading part numbers on components to inspecting printed products before shipment.

Visual information gathering in a more general sense is also possible with artificial vision systems. Intelligent manufacturing systems often have the need to determine information about a part or component before it is further processed. Visual part identification enables systems to handle a wide variety of components without the extensive implementation of bar code readers and material tracking systems, because identification information is inherent in a part's appearance. Determining the orientation of randomly placed parts is also needed for robot handling, and sometimes the visual determination of location is desirable as well.

Another category of machine vision tasks is that of robot guidance. Vision systems can provide data on the location of certain visible features to aid a robot in performing its task. Systems have been used to locate holes for the automatic insertion of fasteners, to locate randomly oriented parts in a bin for robot picking, and to find seams or edges so a robot could be made to follow them while welding or applying adhesive. As robots become fully mobile, machine vision systems will also provide valuable input to allow for autonomous vehicle navigation. Visual input would allow mobile robots to navigate successfully in complex and dynamic environments.

Inspection, which is the subject of this book, is the oldest category of machine vision task, and it presently represents the widest variety of installed systems. Visual inspection systems are typically used to verify the correctness or completeness of manufacturing operations, and they are usually used to supply pass/fail information. Machine vision-based inspection systems are typically more repeatable than human inspectors, and they can often inspect products at a far higher rate. Visual measurement or grading systems are a natural progression from pass/fail inspection systems. Rather than simply providing pass/fail information, these systems provide numerical data about the size of a product or its features. These systems can grade products into various categories, provide input for statistical process monitoring, or generate feedback information for on-line control of product dimensions.

2.2 DEFINITION OF VISUAL INSPECTION

The customer for a product is the final judge of its quality, and the purpose of an inspection is to determine if a product will meet customer expectations. Whether the process delivers a product to a final customer or to an internal customer (e.g. the next process on an assembly line), the inspection procedure and the specifications or features that it is designed to verify or detect must represent the needs and desires of that customer. Keeping internal standards and specifications in line with customer expectations can only be accomplished through a good relationship and regular communication with the customer.

In order to maintain this customer focus, I favor the use of terminology that refers to the inspection of a **product**, rather than more generic terms like **item**, **part**, or **object**. The use of the term **product** emphasizes the idea that something is being produced for a customer, and that the purpose of the inspection is to ensure that the product meets the customer's expectations.

It is also important to make explicit some of the ideas that are implicit in the term **inspection**. Inspection implies a procedure that does not alter the product in any significant way, as opposed to functional or stress testing in which the product is exercised in some manner. This book discusses visual inspection using visible light, but inspections can also be carried out using other forms of radiation like ultra-violet (UV), infra-red (IR), X-ray, and ultrasound because of particular advantages that they may have. UV radiation has been used to inspect dairy products, since their fluorescence characteristics indicate their freshness (Chan *et al.*, 1990). IR radiation has been used to inspect photographic film that is obviously sensitive to visible light (Hollingum, 1984). X-rays have been used to inspect drill holes in metal (Hedengren, 1986; Mundy *et al.*, 1992; Noble *et al.*, 1992), and ultrasound has been used to inspect hardboard and plywood for internal defects. Many of the techniques that will be discussed here are applicable to images generated by these alternative methods of gathering visual information.

Another common form of inspection relies on touch probes and is conducted by devices such as coordinate-measuring machines. This form and other mechanical-contact inspections are beyond the scope of this book.

2.3 OBJECTIVES OF VISUAL INSPECTION

2.3.1 Introduction

Inspections are usually conducted for two purposes, to make decisions regarding the suitability of the product for customer use, and to gather data on the performance of the product or process. In order to meet the first objective we must somehow define product suitability. Although it has already been said that this definition must reflect the expectations of the customer, the form that this definition should take has not been specified. The most common approach to defining product suitability is to develop some kind of **reference model** for the product.

2.3.2 Reference model inspections for decision making

Reference models for inspection purposes may have both quantitative and qualitative elements (Mair, 1988). One form of quantitative model is often called a **golden template**, a pattern from which the product can deviate only slightly. One example of a product that could be inspected against a golden template is an automotive gasket. If the gasket is made from the proper material, and it matches the reference template closely enough, it is likely that it has been manufactured properly. Golden template models are best used to represent simple products, such as gaskets, that are essentially two-dimensional. For example, template matching has been used to inspect the traces on unpopulated printed circuit boards (Noble *et al.*, 1992; Silven *et al.*, 1989).

Another quantitative form of model is a structure of geometric size, shape, and tolerance specifications. Many machined, cast, molded, or formed products must meet these kinds of specifications. The most common type of this form of model is a mechanical drawing which includes tolerance information. Tolerances are typically specified for attributes like straightness, surface flatness, circularity, profile shape and feature position (Kennedy, Hoffman and Bond, 1987).

Computer-aided design (CAD) systems for product design can also provide geometric reference models against which those products can be inspected. CAD models are particularly suitable as reference models because they are often already in existence, they usually contain tolerance information, and they are exact three-dimensional representations of objects (Newman and Jain, 1995). Because of these advantages, several CAD-based inspection systems have been developed (Dahle, Sommerfelt, and McLeod, 1990; Newman, 1993; Newman and Jain, 1994; Park and Tou, 1990; Silven *et al.*, 1989; Sobh *et al.*, 1993; Takagi *et al.*, 1990; Yoda *et al.*, 1988). Other quantitative models include more general specifications such as weight, volume, hardness, elasticity or chemical composition. Although many of these attributes cannot be inspected visually, some can be linked to visible features.

Models with qualitative elements are more difficult to specify and interpret. Qualities such as appearance are very important for many consumer goods including agricultural and manufactured food products. Appearance is very difficult to inspect with automated systems, since some irregularity is acceptable in many cases. Every chocolate-chip cookie coming off of the line does not have to be identical, but they must meet some unwritten criteria. Consumers even have a difficult time explaining exactly what they are looking for, and sometimes respond with statements like 'I know a bad one when I see one'. Agricultural products present similar difficulties. Attributes that may be perceived as defects by customers, but which are difficult to quantify, include irregular shape, discoloration and bruising.

The finish on the surface of a product is also difficult to judge in an automated system. As with the cookies, each example of a manufactured product need not be finished identically, but some standard of uniformity must be achieved. Any irregularity in pattern, color, texture, or gloss on a surface can be perceived as a defect by a customer. Consider, for example, sheets of yellow paper. If the color differs slightly between sheets in different packages, it is unlikely that this variation will be noticed. If variation occurs on one sheet, however, in the form of blotches, streaks, or specks, a customer is very likely to perceive this variation as a defect in the product.

2.3.3 Inspections for data gathering

The second objective of visual inspection systems is to gather data about the performance of a product or process. Defect identification and counting are two effective means of accomplishing this. Properly identifying specific defects can often lead to the discovery of their cause, and defect counts can reveal the severity of that cause. Defect counts can also be used to judge the overall performance of a process.

The location of specific defects on the product can also provide valuable information. In the production of silicon wafers, for example, a crystal of amorphous silicon is grown from the center outward. When this crystal is cut into wafers, the distance that any defect in the wafer lies from the center is an indication of when the defect was formed during the crystal growth process. This information can be used to track down problems in the process and reduce future defects.

Detailed defect records can also be very helpful when servicing customer complaints. Product failures can often be correlated with defects found when manufacturing the product, and cause–effect relationships can be developed that will lead to more effective inspections and eventually to the production of fewer defective products.

Inspection data can also provide process performance information in a more general sense. Process yields or downfall rates are used as a measure of process performance in many manufacturing industries. Process yield is often expressed as the percentage of product that passes inspection, and the downfall rate is usually the opposite. Information from inspections can also be used to determine the capability of a process for producing a certain level of quality. A statistical analysis of defect occurrences under various process conditions can be conducted with the goal of determining the theoretical minimum defect level achievable.

2.4 INSPECTION ERRORS

Because decisions are made based on inspections, there is always the probability of error. Errors are typically grouped into two types, which statisticians have imaginatively labeled Type I and Type II. In the context of inspection, a Type I error (sometimes called a **false positive**) occurs when a product is identified as defective when in fact it is not. A Type II error (sometimes called a **false negative**) occurs when a product passes the inspection but it is actually defective. Figure 2.1 illustrates the four possible outcomes of a decision based on an inspection, and the situations where Type I and Type II errors occur.

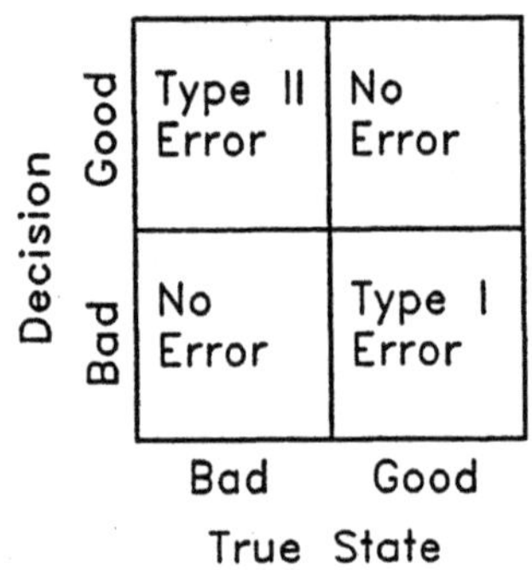

Figure 2.1 Four possible outcomes of an inspection decision.

Often, the quantities of each of these errors are inversely related, that is reducing the number of Type II errors often increases the number of Type I errors. This is true, for example, when a certain specification is simply tightened up in order to solve a quality problem, rather than correcting the root cause of the variation. Fewer bad products are shipped, but at the cost of a higher rejection rate.

2.5 CATEGORIES OF INSPECTION

2.5.1 Introduction

There have been many attempts to dichotomize or categorize the various inspection techniques (Newman and Jain, 1995; Wetherill, 1969). Inspection categories have been created based on the methodology used, the types of decisions to be made, the approach used, and the production area in which the inspection takes place.

2.5.2 Explicit and implicit inspection

One method of inspection involves comparing a product with a reference model to see if it meets the specifications of the model. This method, sometimes called **explicit inspection**, is characterized by the fact that the inspection system knows what it is looking for, and it typically only looks at those features that are expected and represented in the model. Consider the simple drilled block shown in Fig. 2.2. A specification may include tolerances on the length, width, and height of the block, the size and location of the hole, the straightness of the hole, and the angle of the hole to the surface. A visual inspection scheme could be devised to measure all of these variables and determine if a particular block meets the specifications. This model would not, however, detect any unexpected features like an additional hole or a surface discoloration. This approach may seem naive, but the majority of inspection systems in use today employ this strategy.

A second, more thorough method can be used to find novel or unexpected features. First, the product can be inspected for the expected features represented in the reference model, and that analysis can be followed by an inspection that considers the model of the product and searches for any unexpected features. For many manufactured products (paper, aluminum foil and glass are three examples) the reference model consists of a featureless surface. In these situations any visible feature is considered a defect and should be noted during the inspection process.

Humans are very good at this type of inspection, which Lee (1989) has termed **implicit inspection**. An example of implicit inspection is the identification of a product as obviously defective, even when the particular defect has never been seen before. This method of inspection is not employed as widely in automated systems because of the difficulties involved. The fundamental problem stems from the fact that the system does not know what it is looking for, and it must be robust enough to handle a wide variety of features. In addition to just noticing defects it would be very useful if new defects found could be classified into categories based on their similarity to known defects, or if new classifications could be automatically created for sufficiently novel

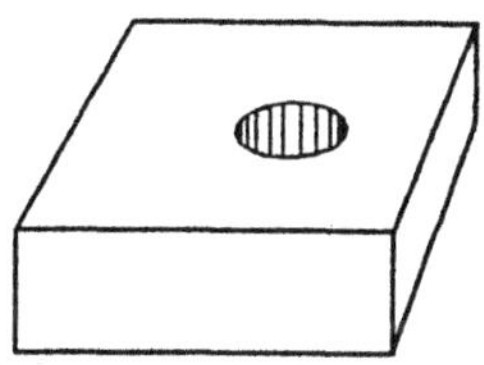

Figure 2.2 Simple drilled block.

defects. As will be seen in later chapters, the application of artificial intelligence and neural networks to inspection can lead to systems that exhibit some of these advantageous traits.

2.5.3 Pass/fail and grading inspections

Inspections can also be categorized based on the outcomes generated. Pass/fail inspections are very common, and are used to determine only if a product is acceptable or not. Grading inspections, on the other hand, are used to sort products into categories based on their quality or suitability for use. Grading inspections usually generate quantitative information that is used in making the grading decision (Wetherill, 1969).

2.5.4 Statistical sampling and 100% inspection

There have traditionally been two different approaches to inspection, one where each and every product is inspected (100% inspection) and another where only a small sample of the products produced is inspected (statistical sampling). Statistical sampling judges an entire batch of products based on a small sample population drawn from the batch. The sample is taken in such a way as to be statistically representative of the batch. The entire batch is downgraded, reworked or scrapped, based on the performance of the sample. The obvious advantage to this approach is the dramatic reduction in the number of products that must be inspected, saving both time and money. There are two big disadvantages of this approach, however. First, the difficulty of getting a truly representative sample. Without a representative sample, decisions made based on the sample population may not be valid for the entire batch. A sample that is unrepresentative of the general population in some way increases the likelihood of Type I and Type II errors. Although the statistical probability of each error may be small, it is usually non-zero and often represents a large number of products. The second disadvantage is that any decisions made apply to the entire batch, so in the event of a failed inspection the entire batch must be scrapped or reworked, even though it is almost certain that a large percentage of the products in the batch are good.

When 100% inspection is employed, each individual product is inspected. This approach has the advantage of judging each product individually, so decisions made affect only that product. It also has the advantage of guaranteeing the quality of each and every product produced, and this is a necessity for many products. 100% inspection is necessary for products in which a defect can cause serious risk to health or safety. Pharmaceutical labels, aircraft parts, and automotive brake systems are some examples of products that demand 100% inspection. One bad brake shoe is one too many if it happens to find its way onto your car.

The recent phenomenon of hysteric negative product publicity has also made 100% inspection very desirable for many consumer products. A single well publicized incident of product contamination can have dramatic and far-reaching negative effects on a company. When consumer confidence in a product is lost due to such an incident the company often loses considerable market share. Many times the product or its packaging must be redesigned in order to regain lost customers.

2.6 AREAS OF APPLICATION FOR VISUAL INSPECTION

2.6.1 Introduction

In the past, the simple model shown in Fig. 2.3 has often been used to represent a procedure or operation. Whether the procedure involves writing a letter, making a meal, or building an automobile, there are inputs, some process, and some resulting output. This model has recently been updated, however, as shown in Fig. 2.4. Although only the labels have changed, these labels carry significant meanings. The inputs for the process come from a supplier, and this implies a relationship with the supplier. Many companies have found that developing positive relationships with their suppliers leads to better overall performance.

Even more important is the realization that any process has a customer. If there is no customer, there is most likely no reason to perform the process in the first place. This model is also very scalable, as it can represent an entire company, with many suppliers and customers, or it can represent one very small step in the assembly of a single product. In the latter case the suppliers and customers are internal, but

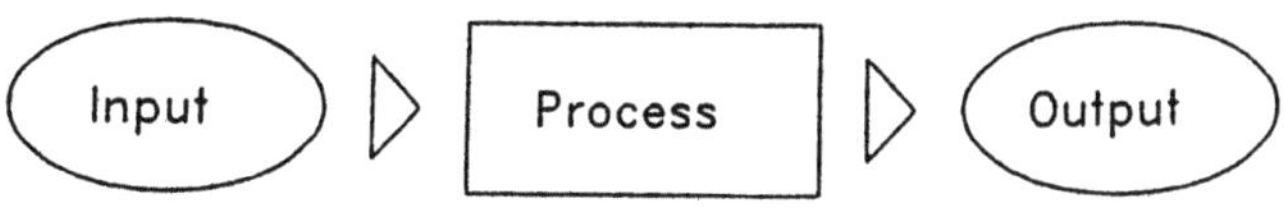

Figure 2.3 Traditional process diagram.

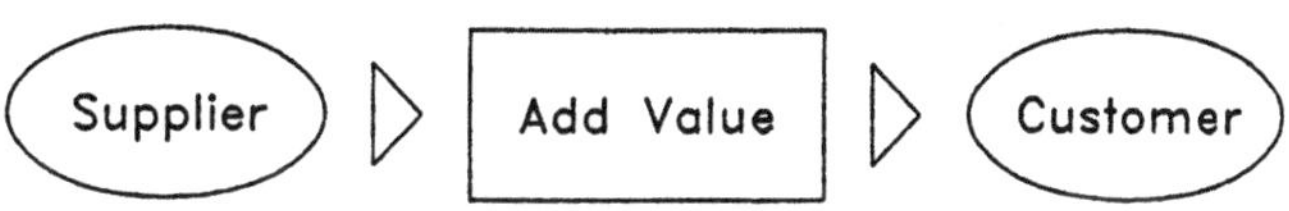

Figure 2.4 Modern process diagram.

they are there, and relationships with them are equally important. Furthermore, each process performed in a company should add value to the product. The value added is ultimately determined by the final customer, the one that pays for the product. Because most procedures can be represented by the supplier–process–customer model, inspections are usually applied in three areas: incoming inspection, in-process inspection, and final inspection (Dorf, 1988; Kennedy, Hoffman and Bond, 1987; Tarbox and Gerhardt, 1989; Wetherill, 1969).

2.6.2 Incoming inspection

Inspection of incoming goods is usually used to verify the quality or suitability of raw materials, components, or subassemblies. Incoming inspections are usually pass/fail, with goods being judged as acceptable for further processing or rejected as unacceptable. Inspections are also occasionally used to grade incoming goods, with goods of differing grades going into different processes or product lines. Incoming inspections are also used to gather data to be used in evaluating certain materials or suppliers, and to provide feedback to suppliers on the performance of their products.

2.6.3 In-process inspection

In-process inspections are typically used for other purposes. Inspections are often required to verify that some process or operation has been completed properly, so the product can move on to the next stage of production. In-process inspections are also used to judge the quality of a product as it is being produced, so that defective products can be removed or reworked as early as possible in the production cycle. This can provide considerable cost savings. For example, one study found that correcting a flaw in a bare printed circuit board had a cost of 25 cents, whereas repairing the same flawed board once it was completed had a cost of $40.00 (Zuech, 1988). In-process inspections are also used for safety reasons, for example to verify that all plugs and seals in a vessel are in place before it is pressurized. In-process inspections can also be used to collect valuable data on the type, number and location of defects. This information can be fed back to the process to help find the root cause of defects and eventually eliminate them. It can also be fed forward to subsequent processes so that defective areas of the product can be avoided. One example of this is paper production, where the slitting of the sheet into narrow rolls can be adjusted to minimize the waste caused by defective areas in the sheet.

2.6.4 Final inspection

Final inspections are probably the type that we are most familiar with. They are conducted immediately before a product is shipped, with the goal of determining if the product will meet customer expectations. Again, both pass/fail and grading inspections are used. Pass/fail inspections are most common, determining whether or not a product is suitable for customer use. Failing products are usually either scrapped or reworked. Grading inspections are sometimes used to determine what market a product will be sold to. Top grade products are sold to the primary market, whereas lower grade products are sold to secondary markets, sometimes under a different brand name or marked as 'seconds' or 'irregulars', or to a controlled group such as company employees. Products not suitable for any market are scrapped or reworked. Grading inspections are also used to classify agricultural products such as eggs and meat into different market categories, reflecting consumer preference more than product quality.

2.7 BENEFITS OF AUTOMATED INSPECTION

2.7.1 Introduction

Early automated inspection systems were installed primarily as labor-saving devices. Inspections and measurements that were done by humans were converted to machines, saving labor hours. Soon, however, other benefits became apparent. As the capabilities of automated visual inspection systems improved, the speed and accuracy attainable exceeded those of human inspectors. The application of automated visual inspection systems allowed companies to produce higher quality goods at higher speeds, with a lower cost and without exposing workers to boring, repetitive and sometimes dangerous inspection tasks.

2.7.2 Lower cost

Because it is concrete and measurable, the favorite justification for any improvement in a manufacturing plant is cost savings. The cost of human visual inspection has been estimated to account for at least 10% of the labor cost for manufactured products (Mair, 1988). Although many automated inspection systems have a high initial cost, the cost of operation is very low. These systems work very reliably with minimal down-time, and they can significantly reduce the labor required to produce a high-quality product. In-process inspection also allows defective products to be removed from the production line as soon as possible, resulting in savings in material, time, energy and labor that would be expended in subsequent processing (Harnarine and Mahabir, 1990; Skaggs, 1983).

2.7.3 Higher speed

Visual inspection creates a bottleneck in many production systems. Humans can only perform even the simplest of visual inspections at a rate of a few per second (Skaggs, 1983), and complicated inspections can take several minutes. The speed required for performing inspections is usually related to the complexity of the product. Complex products are more time consuming to produce, so they typically allow more time for inspection. Inspection times for complicated printed circuit boards and integrated circuits, for example, can be several minutes (Winkler, 1983). Simple and inexpensive products, however, must often be inspected at a rate of several per second. Fortunately, automated systems have been designed to handle these tasks. Automated visual systems have graded oranges at 64 per second (D'Agostino, 1988), sorted shrimp at 20 per second (Kassler, Corke and Wong, 1991), and inspected cigarette packages at 8 per second (White *et al.*, 1990), glass bottles at 16 per second (Novini, 1990), and metal beverage containers at 10 per second (Novini, 1990).

Reductions in inspection time through the application of automated systems have made it possible to consider 100% inspection in many cases where that was previously thought impossible. High-speed systems have made 100% inspection feasible for inexpensive products like fluorescent lamps (Bains and David, 1990), screws (Olympieff, Pineda and Horaud, 1982), rivets (Batchelor and Braggins, 1992), and plastic bottle caps (Laligant, Truchetet and Fauvet, 1993).

2.7.4 Improved product quality

Producing low quality goods is not profitable. Downgraded products sold to secondary markets are often priced on a break-even basis, so no profit is made. Products scrapped or reworked usually result in an outright loss. Because of this, there is an increased emphasis on total quality today. Many companies have made efforts to minimize or eliminate the production of downgraded products. Making each product right the first time is the goal. Inspection can play an important role in this effort by identifying defects early in the production process where they can be corrected at minimal cost, and providing defect information that can be used to track down the root cause of quality problems.

2.7.5 Accuracy and consistency

Human operators suffer from boredom and fatigue when performing repetitive tasks such as inspection, and it has been shown that humans performing 100% inspection of products are actually only about 80% effective (Smith, 1993). Achieving true 100% inspection with human inspectors can only be accomplished by having inspections performed redundantly, which further increases the cost (Dreyfus, 1989). Humans are also influenced by other factors, such as becoming inconsistent as production speeds increase or altering their inspection standards based on real or imagined quotas (Newman and Jain, 1995). Automated visual inspection is often the only way to accomplish consistent, accurate, high speed inspection of all products at production line speeds (Harms, 1992).

2.7.6 Improved worker safety

Automated visual inspection systems have the additional benefit in that they can operate in hot, noisy or dangerous conditions that human inspectors could not tolerate. Worker safety can be improved when dangerous vision-oriented inspection tasks can be conducted by machines, removing people from hazardous areas. Tokyo Electric Power Company, for example, has developed an automated system for inspecting pipes for water and steam leaks in a high-radiation area in a nuclear power plant (Yamamoto, 1992).

2.8 COMMERCIALLY APPLIED INSPECTION SYSTEMS

A designer of an automated inspection system must be familiar with inspection system hardware and software, but an intimate understanding of the inspection problem being addressed is even more important. Due to this specific nature of visual inspection system design, the companies supplying commercially available automated systems have typically become specialized in narrow areas. A company that supplies inspection systems for the integrated circuit electronics industry would not necessarily be a good candidate to supply a system for inspecting automotive body panels. In a recent survey article, Newman and Jain (1995) identified the following inspection system providers (and products inspected): GEI (semiconductor packages), IRT (solder joints), Four Pi (solder joints), KLA Instruments (printed circuit boards, silicon wafers), Machine Vision International (surface finish), European Electronic Systems (metal surfaces), and Computer Recognition Systems (oranges). Some companies providing more general purpose systems are Adept, Allen-Bradley, Acuity Imaging and Itran.

These general-purpose systems will typically require more development effort on the part of the user, since they must be adapted to the particular application.

Aside from these commercially available systems, several companies have had sufficient interest in automated visual inspection to develop their own systems for internal use. Delco has developed an inspection system for checking chip structure (Miller, 1985), General Motors for openings in car bodies (Rushlow, 1983), I.B.M. for keyboard springs (Moir, 1989), Mitsubishi for television picture tubes (Kishi, Hibara and Nakano, 1993), Saab for flywheel castings (Van Gool, Wambacq and Oosterlinck, 1991), Texas Instruments for integrated circuit packages (Narasimhan, 1992), Volkswagon for automobile body dimensions (Kurowski, 1988), and Westinghouse for turbine blades (Miller, 1985). The size of each of these companies is an indication of the commitment in resources required to develop an automated visual inspection system from scratch.

In addition to those already mentioned, several applications of automated visual inspection systems are briefly described in Appendix A.

2.9 LIMITATIONS OF AUTOMATED VISUAL INSPECTION

When compared to the manufacturing process itself, the automation of visual inspection has come fairly slowly (Newman and Jain, 1995). Although many successful applications of automated visual inspection have been listed in this chapter, there are several factors that currently limit the feasibility of such systems.

First, the design of automated visual inspection systems is a difficult task. Knowledge in many technical areas is required, including illumination, cameras, computer interfacing, programming and image processing. Specific knowledge of the inspection problem to be solved is also required, and finding or developing all of this expertise in one person or a small group of people is difficult. Initial system cost is also a limiting factor to the broad application of automated visual inspection. Simple systems cost around \$30 000, and complex systems can easily exceed \$200 000 in cost (Newman and Jain, 1995). The speed of automated systems has also been a limiting factor in the past, but recent dramatic increases in computational speeds have made very high speed inspection systems feasible. Another disadvantage of current inspection systems stems from the fact that their installation often requires undesirable or inconvenient process changes. Products sometimes have to be stopped, or separated, or specially illuminated in order to be inspected successfully.

Although many of the shortcomings just mentioned are being

successfully addressed, there is one trait that automated inspection systems currently lack and which will be more difficult to develop, and this is best described as **flexibility**. Current systems are typically each custom-designed for a specific task. In a recent article surveying well over 100 applications of automated visual inspection, Newman and Jain (1995) cited only one experimental system that was designed for more general-purpose visual inspection (Marshall, 1990). Flexibility will not only lead to wider application of systems, it will also allow systems to be more robust in dealing with problems such as objects in varying positions and orientations, overlapped objects and varying levels of illumination.

Human inspectors perform very well in inspection situations that demand such flexibility, because in addition to excellent visual capabilities, many inspection tasks require a fairly high level of reasoning ability on the part of the inspector. A higher level of artificial intelligence must be incorporated into automated inspection systems if they are to deal robustly with the aforementioned problems. Intelligence is also required to deal with novel and unexpected situations and to learn from past experience, two capabilities that humans have that would be very desirable in automated systems.

2.10 RESEARCH GOALS

The goal of the research that is described in this book was to address some of the weaknesses of current visual inspection systems by developing a prototype for an intelligent visual inspection system. In doing so, it was desired to take advantage of new theoretical developments in the fields of vision and artificial intelligence. Some examples of these developments include the application of massively parallel neural network-like processing rather than traditional image processing, and the incorporation of artificial neural networks for product and defect recognition. It was also desired to incorporate some form of learning into the system, so that it could adapt to changing situations without reprogramming.

Although this system was intended for eventual application in manufacturing environments, it was to be different from almost all previous systems employed in manufacturing. First, the emphasis was on two- and three-dimensional object recognition rather than measurement or gauging. Second, and most important, an effort was made to assure that this system would be flexible, i.e. applicable to a wide variety of tasks under diverse circumstances. Nearly all existing vision systems for manufacturing applications are restricted to a very specific problem domain and require rigid conditions in which to operate. This effort sought to create a more generic, general-purpose system. Finally, the realities of manufacturing applications were taken into account, so that the resulting system would be practical to implement for a realistic cost, and so it would be capable of performing its function in a reasonable amount of time.

Part Two

Fundamentals of Artificial Vision Systems

3

Biological vision systems

3.1 INTRODUCTION

If the eventual goal of an automated inspection system is to duplicate human visual capabilities, those capabilities must first be well understood. Since the early 1900s researchers have attempted to define the limits of human visual and perceptual capabilities. These efforts have included the destructive dissection of the biological structures responsible for vision, the direct measurement of the responses of brain cells in animals to visual stimuli, and psychological testing of humans in which subjects are shown a specific visual stimulus and are then asked to explain their perception of that stimulus. Although much of this research is not recent, it forms the basis of most artificial vision systems presently under development.

The intent of this chapter is to review the body of physiological and psychological research relating to human vision, and it consists of two main parts. The first part presents physiological facts gathered from direct experimentation. The nature of light and how it is sensed by biological structures are discussed, and the characteristics of those structures are presented, with particular emphasis on the mammalian retina. The current state of experimental research into the composition and function of these structures is also reviewed. The second part of this chapter presents the results of studies employing the psychological approach to visual interpretation. The characteristics of human perception, the Gestalt laws of visual organization, and the visual properties of objects are reviewed. Finally, the visual properties to which human perception exhibits a high degree of invariance are emphasized, because of their importance to object recognition.

3.2 BASIC PHYSIOLOGY OF VISION SYSTEMS

3.2.1. Nature of the input

Light, in its most basic form, is electromagnetic (EM) radiation. Electromagnetic radiation is usually characterized by two models, the wave model and the photon model. Each model serves to explain some aspects of the behavior of EM radiation, and no theory unifying the two models into a common description has yet been agreed upon. Using the wave model, EM radiation possesses four characteristics, namely wavelength, phase, amplitude and propagation velocity. The wavelength is the physical measurement of the distance between peaks in the EM waveform (Fig. 3.1). The EM spectrum ranges in wavelength from 10^{-1} to 10^{11} nanometers (nm), going from X-rays to ultra-violet, visible light, infra-red, microwaves and radio waves. Visible light represents only a very small portion of the EM spectrum (Fig. 3.2). In the visible light region, wavelength is perceived by human beings as color. Amplitude represents the height of the peaks in the waveform (Fig. 3.1), and is perceived by humans as intensity.

Phase, usually represented as an angular displacement in the sinusoidal waveform, represents the displacement of the wave from some point of reference (Fig. 3.1). Although the human perception of phase is not clear, it seems to provide a clue as to the relative depth of objects. The same light source reflecting from two objects slightly displaced in depth with respect to the viewer will result in two waveforms at the viewer that are slightly out of phase with each other. Although this phase shift does not seem to be directly perceived, its effects are very noticeable. A standard color photograph reproduces the wavelength and intensity information from the original scene, but it loses the phase information, and it is perceived as two-dimensional. A laser hologram, however, reproduces the intensity and phase information but loses the wavelength information. For this reason, it is perceived as three-dimensional but monochromatic.

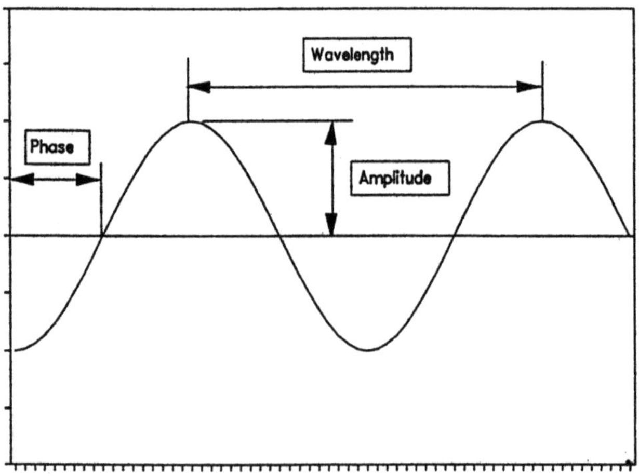

Figure 3.1 Electromagnetic waveform.

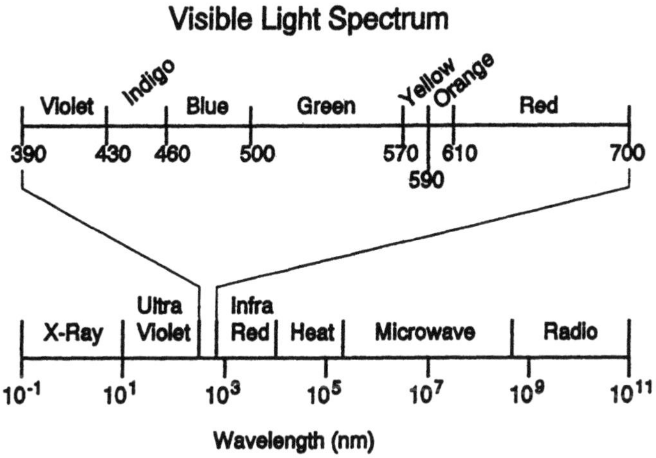

Figure 3.2 Electromagnetic spectrum.

Propagation velocity represents the speed at which the light wave peaks propagate through space. Propagation velocity is determined by the optical density of the medium through which the light is passing, and it reaches a maximum of 3×10^8 m s^{-1} in a perfect vacuum. Although propagation velocity is not perceived directly by humans, the change in velocity as light passes through different media has obvious and important effects.

Visible light interacts with the physical environment in several ways (Halliday and Resnick, 1974). Absorption occurs when light strikes matter, and the energy possessed by the light wave is absorbed by the matter. When this happens, the light 'disappears'. If an object absorbed all of the light falling upon it, however, it would not be perceived as invisible. Instead, the complete absence of light would be perceived, resulting in a totally black void.

Diffraction of light occurs when light strikes particles and is scattered. Different wavelengths of light are diffracted to a different degree, with shorter wavelengths being scattered the most. This is the reason that the sky is perceived as blue. Since the blue end of the spectrum is diffracted more, more blue light reaches the ground. If the Earth had no atmosphere to diffract incoming light, the sky would appear black unless the sun was directly in the field of view. This situation exists on the moon.

Refraction occurs when light waves are bent due to the previously mentioned effect of optically dense materials on the propagation velocity. This is the reason why underwater objects, when viewed from above the surface, appear to be displaced. This is also the reason why lenses, devices of obvious importance to vision, work the way they do.

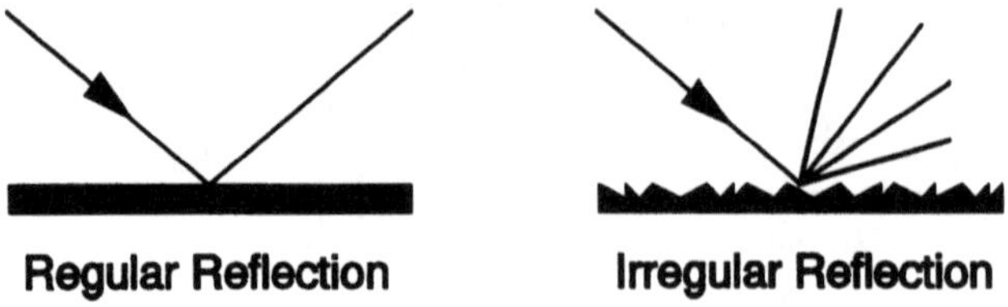

Figure 3.3 Types of reflection.

Reflection occurs when lightwaves encounter objects and bounce off. There are two types of reflection, regular and irregular (Fig. 3.3). Regular reflection occurs when light encounters a smooth surface such as a polished mirror and the light is reflected in an organized manner. Irregular reflection occurs when light strikes a textured surface and is reflected in a disorganized manner in many directions. Usually, when reflection occurs, some of the incident light is also absorbed. Furthermore, the wavelength mix of the absorbed and reflected light differ, and this determines how the color of a surface is perceived. Leaves absorb more of the red end of the spectrum than they reflect, so they are perceived as green.

Although light is virtually everywhere in the daylight environment, an animal or person can only detect those rays that converge at the point where it can be sensed. Gibson (1966) coined the term **ambient optic array** to describe the pattern of intensities and wavelengths present at the specific sensing point. The word **vision** describes the analysis of the information available in the spatial and temporal pattern of visible light in the ambient optic array.

3.2.2 Biological light-sensitive structures

Many biological molecules absorb EM radiation and change chemically as a result. Photosynthesis is a common example of this, where chlorophyll molecules absorb light and produce sugars. Light incident on plants also triggers other responses, like the turning of leaves toward a light source. If a tree is subjected only to light from a single non-moving source, and a fixed pattern of shadow is present in that source, the leaves of the tree will grow away from the shadow and toward the light, and in this way the pattern of shadow will be reproduced in the array of tree leaves. This represents a crude form of biological vision. Leaves grow toward the light because they see it!

Even single-celled animals exhibit light sensitivity. An amoeba moves by extending its flexible body into new areas. If one of the extensions encounters a bright light source, the organism will not move in that direction (Bruce and Green, 1985). This is another example of crude visual behavior. The amoeba uses cues available in the visual environment to avoid brightly lit areas that are hazardous to its health.

Higher animals have developed cells that are specialized to the light detection task, called **photoreceptor cells**, that convert light energy directly into cellular receptor potentials. Some of these photoreceptors also possess pigments that allow them to be sensitive to only certain wavelengths of light, making color perception possible. For an array of light from the physical environment to be intelligently interpreted, however, it must be well organized. This usually requires an array of photoreceptor cells, and some way of controlling the light falling on the array. Some lower animals have developed just such an array, with an orifice above the array to organize the incoming light. This type of 'eye' works like a pinhole camera, where a small orifice results in a clear image on the photoreceptor array. Unfortunately, the orifice needs to be very small in order to provide good focus, and such a small orifice allows very little light to pass, resulting in poor response to low light intensities. Evolution has solved this problem in higher animals by making the orifice larger, and spanning it with a refracting lens that can focus the image accurately on the photoreceptor array. The human eye is the result of a high degree of such evolution.

In higher mammals the photoreceptor array, known as the **retina**, is composed primarily of two types of cells, called **rods** and **cones** because of their shapes (Uttal, 1981). Rod cells work well in dim light conditions, and respond primarily to intensity differences. Cone cells occur in three basic varieties, with each variety responding to a different region in the wavelength spectrum. These cells are responsible for the perception of color, and are usually referred to as red, green and blue (RGB) cells. The R, G and B cells have a peak response to wavelengths of approximately 620 nm, 540 nm and 440 nm, respectively, as illustrated in Fig. 3.4 (Marks *et al.*, 1964). The rod and cone cells are arranged into an array to form the retina, and it is interesting to note that the cells appear to be installed 'backwards' (Fig. 3.5), so that the light must pass by the cell body in order to reach the photosensitive region. This apparent evolutionary mistake actually results in very high directional sensitivity in the array, and eliminates the effects of internal reflection in the eye (Rodeick, 1973).

The human retina is composed of a total of approximately 120 million rod cells and 7 million cone cells (Bruce and Green, 1985), but the rod and cone cells are not distributed evenly. The central area of the retina called the **fovea**, which is a roughly circular region that subtends about 10° of arc, consists of a highly concentrated region of primarily cone cells. This area provides high visual acuity (i.e. excellent resolution to fine spatial features) in bright light conditions. The area outside of the

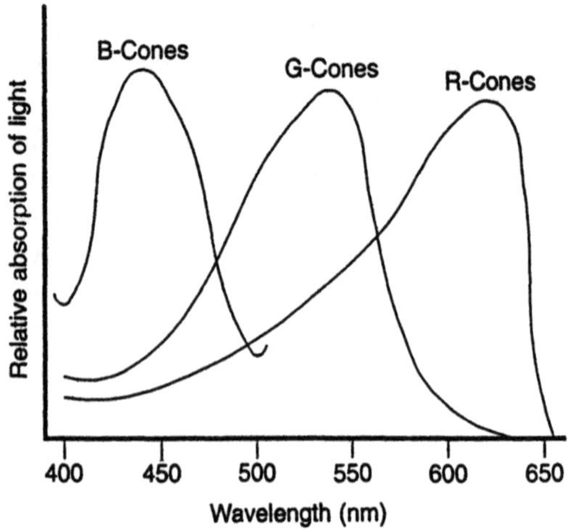

Figure 3.4 Cone cell color responses in mammals.

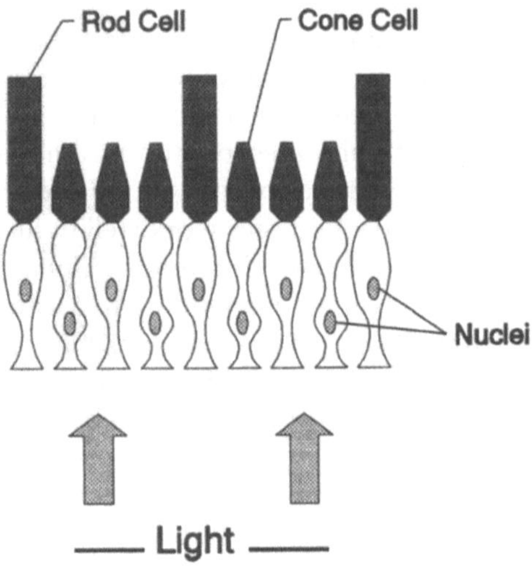

Figure 3.5 Arrangement of receptor cells in the mammalian retina.

fovea called the peripheral area consists primarily of rod cells, is not as highly concentrated, and provides excellent visual sensitivity to very low intensity features in dim light conditions.

In addition to the lens and retina, the human eye is also equipped with an adjustable pupil that can reduce its orifice to aid in focusing in bright conditions, or increase its orifice to allow more light to enter the eye in dim light conditions. The pupil alone, however, does not account for the incredible intensity range of the human eye. The rod and cone cells possess the ability to adapt to the average level of illumination present. This adaptability allows the human eye to respond to a log 7 intensity range, where the brightest light is a factor of 10^7 greater than the dimmest perceivable light.

3.2.3 Low-level feature processing mechanism

Although the receptor cells (rods and cones) of the retina sense visual stimuli, they do not transmit this information directly to the brain. Signals from the receptor cells pass into a network consisting of four types of cells, namely horizontal cells, bipolar cells, amacrine cells and ganglion cells. These cells are organized into two layers called the **outer plexiform layer** and the **inner plexiform layer** (Dowling, 1968). Figure 3.6 shows a schematic diagram representing the nature of the connections between the various types of cells. The horizontal cells provide horizontal connections in the outer plexiform layer, perhaps facilitating the exchange of signals among receptor cells. The amacrine cells appear to perform a similar task in the inner plexiform layer. The bipolar cells form the primary connection between the layers, and they also perform grouping of the signals from the receptors. The connections occur in such a way as to group only small numbers (less than 10) of cone cells in

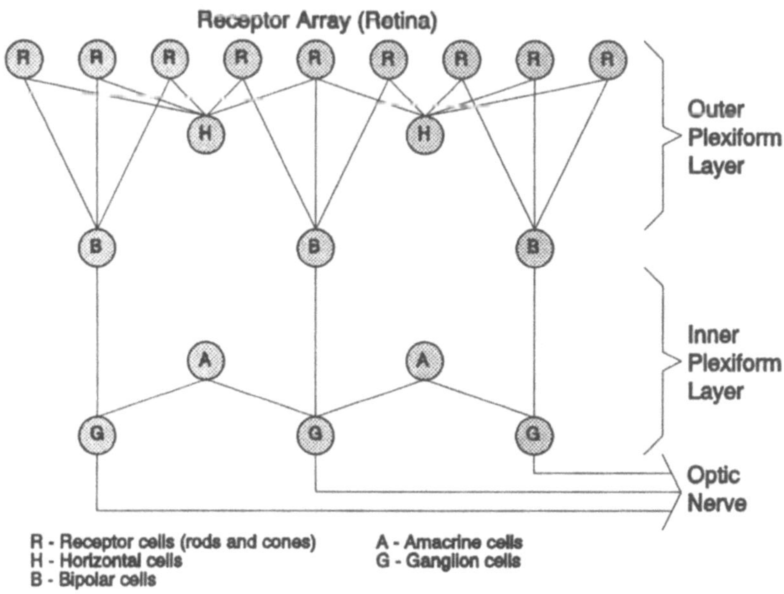

Figure 3.6 Cellular connections of the mammalian retina.

the fovea, while larger numbers of primarily rod cells are grouped in the peripheral region. This is consistent with the performance of these portions of the retina described earlier.

The signals transmitted from the bipolar and amacrine cells connect primarily to the ganglion cells, and the outputs of the ganglion cells form the optic nerve, which forms the connection to the brain. Although the function of the inner and outer plexiform layers is not totally understood, they appear to perform two tasks. First, although there are a total of about 127 million receptor cells in the retina, there are only about one million ganglion cell signal paths in the optic nerve. This represents a data compression of better than 100:1 directly at the retina! Also, it is believed that this network of cells performs some kind of transformation on the image present at the retina, although the nature of this transformation is not well understood.

In order to shed some light on the function of the ganglion cells, tests measuring their response to various stimuli have been conducted. These tests have shown that ganglion cells respond to approximately circular concentric receptive fields on the retina (Kuffler, 1953). Some respond positively to a bright stimulus in the center of their receptive region, and negatively to a bright stimulus in the area surrounding the central area. This is referred to as the on-center–off-surround response. Some respond in exactly the opposite manner, and exhibit what is called the off-center–on-surround response. A mathematical description of these responses has been developed in which the response of the central area is represented by a narrow Gaussian function and the response of the entire receptive region is represented by a broader Gaussian function, as seen in Fig. 3.7. The response of the cell is then represented by the difference between the two Gaussian values, resulting in either an upright or an inverted 'Mexican hat' function. These Gaussians are actually three-dimensional (Fig. 3.8), as they are rotated about their vertical central axis. This is known as the **difference-of-Gaussians** (DoG) model of ganglion cell response (Enroth-Cugell and Robson, 1966).

Ganglion cells have also been found to respond to different colors. As previously discussed, color information is provided by the receptor cells in the form of signals from three different types of cone cells (R, G and B) and intensity information is provided by rod cells. These input signals are resolved into three categories of responses in the ganglion cells. Some cells appear to respond to red and green, with those being subdivided between cells that respond positively to red and negatively to green (+R/−G) and those that respond in just the opposite manner (−R/+G) (Abramov, 1968). Other cells respond to the colors yellow and blue in a similar manner, with some responding positively to yellow and negatively to blue (+Y/−B), while others respond negatively to yellow and positively to blue (−Y/+B). Still other ganglion cells respond

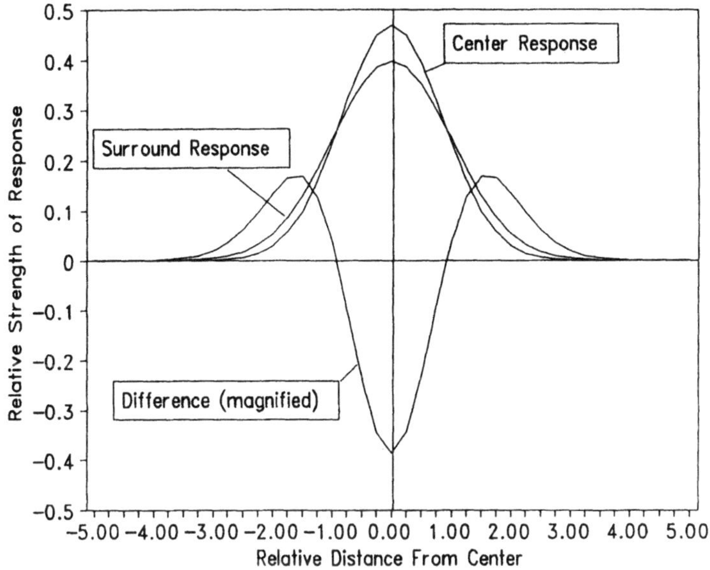

Figure 3.7 Ganglion cell response.

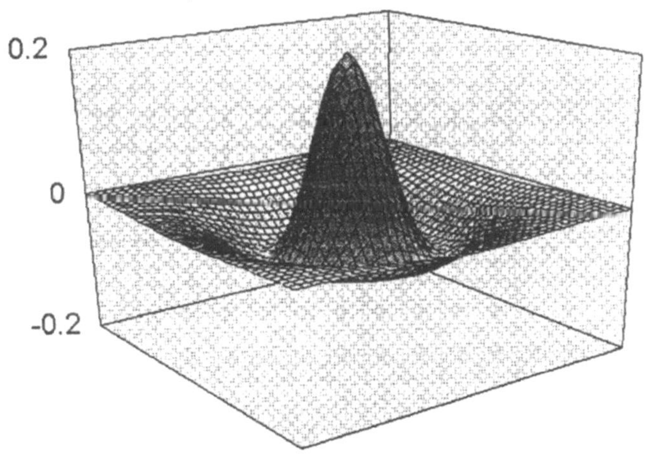

Figure 3.8 Difference-of-Gaussian response model.

primarily to the intensity of the stimuli rather than the color, and these are referred to as black/white (Bk/Wh) cells (De Valois *et al.*, 1958).

The ganglion cell responses can be further subdivided into other categories, the two most prominent being the X-cell response and the Y-cell response (Levine, 1985). Ganglion cells designated as X-cells have a response that is approximately linear to the amount of stimulus present at the concentric receptive region, whereas Y-cells exhibit a transient and non-linear response that appears to be triggered by a change or movement in the stimulus rather than by its absolute value.

Although the complexity and diversity of the inner and outer plexiform layers is impressive, they appear to perform only a rather low-level preprocessing and compression of visual data before it is sent to the brain proper where futher processing is done. The outputs of the ganglion cells, after travelling down the optic nerve, terminate primarily at the lateral geniculate nuclei (LGN) of the thalamus. The LGN provide the interface between the ganglion cells and the primary visual cortex of the brain.

The landmark work of Hubel and Wiesel (1962) provided experimental evidence of the structure and function of the visual cortex of the cat, and their description of its operation has become widely accepted. Their results indicated that the primary visual cortex is arranged into layers, and that each layer is composed of simple and complex cells. Simple cells respond to geometric features such as edge segments, line segments, and corners in a specific position and orientation in the visual field, while complex cells in a given layer respond to the same features as the simple cells, but over a wider variety of positions.

Further work by Hubel and Wiesel (1977) caused them to postulate the existence of **hypercolumns**. Each hypercolumn responds to features in a specific area on the retina, and a hypercolumn consists of approximately 20 layers. Each layer responds to features of a similar orientation, resulting in a resolution of approximately 9° per layer over the 180° orientation range. All of the simple and complex feature detectors (edge, line, corner etc.) are thought to be present in each layer. An interconnected hierarchy of hypercolumns, then, would be capable of responding to moderately complex features at any position or orientation in the visual field.

It is fair to ask how the signals from the concentric receptive regions of the ganglion cells could be used to detect edge and line segments and other features. One possible mechanism for the detection of lines is shown in Fig. 3.9. A single neuron cell in the primary visual cortex could simply sum the inputs from a line of overlapping on-center–off-surround ganglion cells, resulting in a detector that is highly sensitive to lines in a specific orientation (Lindsay and Norman, 1972). The degree of orientation selectivity that is possible with this arrangement is shown by the graph in Fig. 3.10. Adding a parallel line of overlapping off-center–on-surround cells to this arrangement could result in a detector that is sensitive to edges where an intensity or color change occurs with a similar degree of accuracy, and other arrangements can be imagined that would detect ends of lines and corners (Marr and Hildreth, 1980). Although these are not necessarily the mechanisms actually employed,

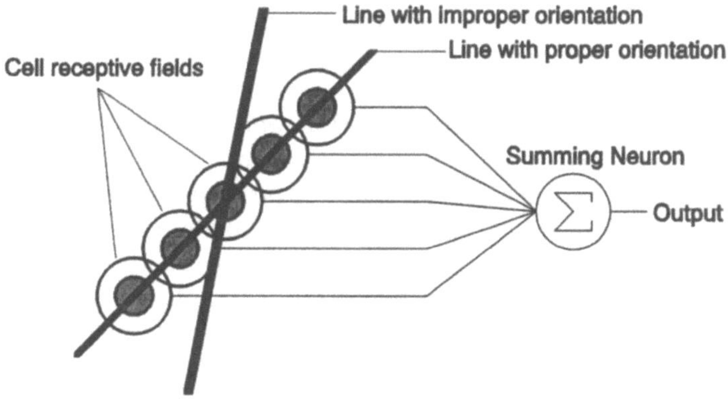

Figure 3.9 A possible cellular line detection mechanism.

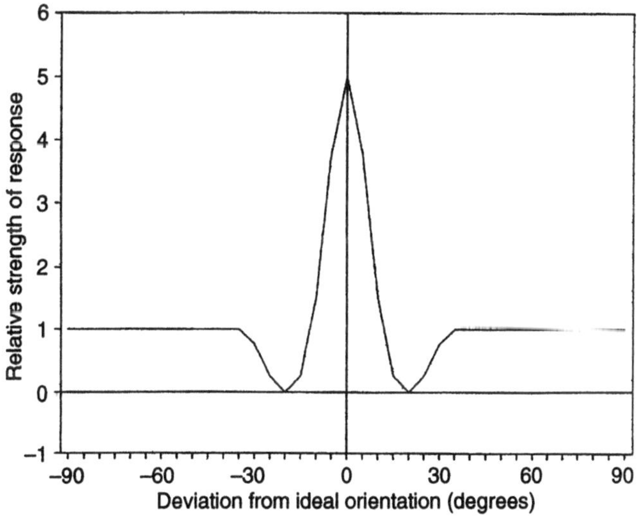

Figure 3.10 Orientation sensitivity of the suggested line detector.

they provide proof that extracting these features from ganglion cell output signals is indeed possible.

3.2.4 Intermediate visual processing mechanisms

The previous section described the possible structure and function of a cellular network that was capable of detecting features such as lines, edges, line ends and corners in the visual array. Other tests have shown that certain cells of the visual cortex are also sensitive to only a specific band of spatial frequencies. These cells would allow features to be extracted at different scales, with high spatial frequencies resulting in very detailed resolution and lower spatial frequencies providing a coarser, more global view. This process of recognition at various resolutions is called **scale-space mapping**, and it is consistent with the human visual recognition process. Humans are thought first to recognize things by their general shape (where your grandfather and a grandfather clock may look similar), and then to verify the identification by proceeding to smaller scales where higher resolution is possible. The process may first differentiate between the shape of a grandfather clock and a human figure, then between a mannequin and a real person, and finally between specific persons, concluding with the identification of your grandfather.

The information available to the mammalian vision system at this stage consists of primitive features detected at various spatial frequencies. The nature of the additional processing that occurs beyond this point, however, is largely unknown. Studies performed on rhesus monkeys (Van Essen, 1979) have resulted in the conclusion that visual information is represented in a series of 'maps' of the visual field. At least five such maps have been identified in the rhesus monkey, four located in the visual cortex (V2, V3, V3A and V4) and one (MT) in the middle temporal area. These maps appear to be very complex spatio-temporal representations, not simple reproductions of the visual field. It is believed that signals from the primary visual cortex (V1) previously described are 'parcelled out' to the various maps in order to conduct complex parallel processing of the information (Zeki, 1978). Tests have shown that the maps vary greatly in their responses to feature orientation, movement, color and spatial frequency. It is thought that these maps could possibly extract more complex features such as bars, blobs, longer lines and curves (Marr and Hildreth, 1980). Tests have also revealed that bidirectional connections exist between virtually all of the map areas, and feedback connections to the primary visual cortex are present as well. These connections enable complex interactions between the maps, the nature of which is poorly understood.

3.2.5 Summary of the physiology of vision

The previous sections have explained that the mammalian vision system consists of a primary sensory array of receptor cells that are sensitive to both color and intensity, followed by a hierarchical information processing system. The cells that transmit information from the retina to the LGN appear to have concentric receptor fields that possess a difference-of-Gaussians response characteristic. At the next stage, the primary visual cortex, simple features such as fixed or moving line and edge segments are detected with an orientation resolution of about 10°.

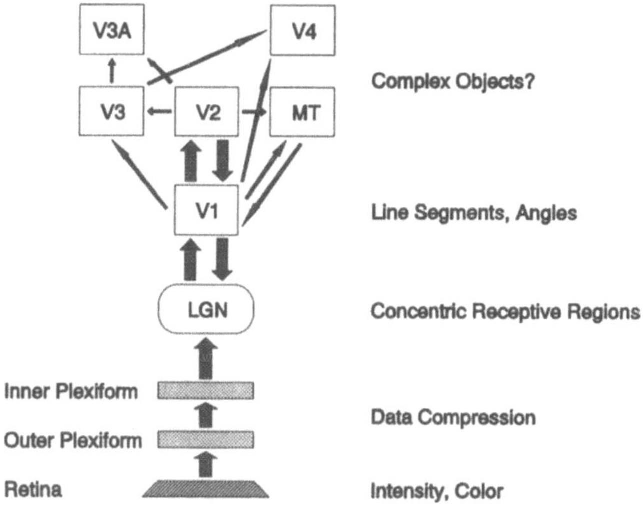

Figure 3.11 Mammalian visual processing hierarchy.

From this point on the information becomes more sparse, but a series of interconnected visual maps seems to exist in which more complex features can be detected with respect to color, orientation, movement and scale. At each stage of the hierarchy, the data is compressed into fewer pieces of information, but each piece of information represents a more complex and abstract entity. Figure 3.11 presents an overall view of the mammalian visual hierarchy.

It should be emphasized that this breakdown represents a rather simplified version of the data available from published research. Many complex features (not well understood) of the visual system have been discovered that are not mentioned here. Also, other theories have been proposed that explain the structure and function of the mammalian vision system rather differently, but an attempt has been made here to integrate the most widely accepted explanations. Finally, direct physical evidence of the processing taking place at the upper levels of the visual hierarchy is sparse or non-existent, so this is where physiology bows out and psychology takes over the task of explaining visual perception.

3.3 HUMAN VISUAL COGNITION

3.3.1 Stages of perception

Psychologists have been testing the visual perception of human beings for several decades. This has essentially been an application of the scientific method to the problem, whereby some theory was proposed which explained one or more aspects of human visual perception and experiments were devised which tested the theory. The results of the experiments either confirmed or refuted the theory, and the theory was modified and the process repeated. The problem with this approach has been that there are many possible hypotheses which explain any given phenomenon, and also many different experiments can be devised to test the validity of these explanations. Therefore it has been impossible to test all possible hypotheses, but many experiments have been conducted nevertheless.

The primary objective of all of these tests has been to explain how a human being, from information assumed to be a two-dimensional representation of edges and lines, arrives at the recognition of specific two- or three-dimensional objects. Some early tests resulted in the division of the process of visual cognition into four stages (Vernon, 1952).

The first stage is called **vague awareness**, and it describes a state in which the observer is aware that 'something is there' but no detail is yet available. Several tests have shown that the presence of light is perceived before any form is resolved (Vernon, 1952). Tests were run in which subjects were exposed to images of very short duration as flashes on a screen in a dark room, and if the exposure was short enough the perception was only of a presence of light. Tests using images of varying intensity also indicated that the presence of light or 'a haze' was detected at very low intensities where form was not detectable. Other experiments were conducted exposing subjects to randomly shaped ink blots, and subjects reported the presence of an object, and the general size and location of the object, before the shape of the object was resolved. All of these tests concluded that this early stage of perception is characterized by a sense of uncertainty. It has been estimated that the vague awareness stage is reached after approximately 1 ms of exposure to a stimulus.

The next stage of perception is called the **generic object** phase (Vernon, 1952). In this stage figures are separated from the background, and the visual field is grouped and organized. Certain objects stand out as being more important than others. Objects are also perceived as belonging to a general category or class (Hake, 1966). It has been estimated that the generic object stage is reached after 10 ms of exposure to a stimulus.

As details of the more important objects in the visual field are resolved, the **specific object** stage is entered. As details are seen, objects are differentiated and identified as specific known objects. It is interesting to note that experiments to determine where in the brain this phase occurs have identified an area very close to the speech center of the brain (Vernon, 1952). This suggests a strong connection between identifying an object and naming the object. In fact, consciously associating a name with an object allows subjects to identify the objects more rapidly. The general success of memory improvement techniques which suggest that a person associates an easy-to-remember mnemonic with a concept that is difficult to remember tends to support this idea. Abstract or unrecognized objects that defy naming are often assigned generic names like 'blob', or are described by naming their subparts, such as 'two lines joined by a curve, with a small square attached to one line' (Vernon, 1952). At the end of the specific object stage of perception, the identification of the object is complete. It is estimated that this phase is completed after approximately 100 ms for simple objects, and it can take considerably longer for complex objects. It has been shown that, in tests involving the identification of aircraft from example pictures, an exposure to the example of 1 s or more was required in order for the image to be adequately retained for later comparison.

The final stage of perception is the **assignment of meaning** phase. In this phase the identified object takes on meaning, and is associated with certain consequences. Questions like 'What is it made of' or 'What is it used for?' or 'Why is it here?' are answered (Vernon, 1952). A sense of familiarity with the object, i.e. the awareness of the presence of the object in one's personal experience, is often described at this phase. The consequences of finding this particular object in the present environment are considered, and potential responses are developed. This stage marks the interface between the more or less automatic identification of objects and conscious reasoning about objects.

3.3.2 Gestalt laws of visual organization

The Gestalt psychologists of the period 1920–1950 concerned themselves with the processes by which line segments, arcs, vertices and blobs are grouped or segmented in order to produce representations of coherent objects. They argued that 'the whole is greater than the sum of its parts', meaning that when human beings view groups of rather simple components as a whole rather than individually, often a more complex object is perceived (Bruce and Green, 1985). In order to explain this phenomenon, the Gestaltists formulated a number of laws of visual organization, involving the following attributes:

1. *Proximity* Objects that are close together tend to get grouped together. This phenomenon is illustrated in Fig. 3.12, where three arrangements of blocks are presented. Because of proximity grouping, the arrangement on the left is generally seen as columns, and the one on the right is seen as rows, while the center arrangement is ambiguous.

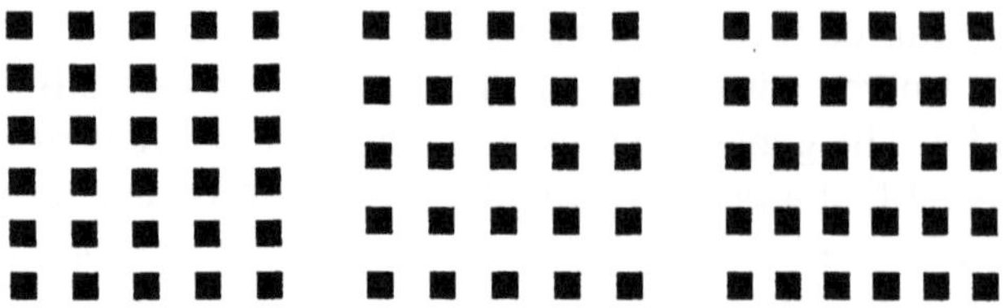

Figure 3.12 Illustration of the human tendency to group by proximity.

Figure 3.13 Illustration of the human tendency to group by similarity.

2. *Similarity* Similar objects tend to be grouped together. This concept is illustrated in Fig. 3.13. The arrangement shown is generally perceived as columns, even though proximity would tend to group the boxes into rows. In this case similarity overrides proximity.
3. *Common movement* Objects that move together are grouped together. A simple experiment was conducted by Gibson and others (1959) to illustrate this phenomenon. Powder was spread on two different pieces of glass, and light was projected through the glass for viewing. Subjects initially perceived a single collection of powder, but when the two pieces of glass were moved relative to each other the subjects immediately perceived two separate groups of powder. A more familiar example of this phenomenon involves camouflage in animals. When a well-camouflaged animal remains motionless it is very difficult to detect, but when the animal moves relative to the background it is almost immediately seen and identified.
4. *Continuation* Objects are grouped so as to preserve smooth continuity rather than to create abrupt discontinuity. This phenomenon is illustrated in Fig. 3.14, where the figure on the left is usually perceived as two smoothly curving line segments rather than two separate V-shaped objects (right).
5. *Closure* When more than one interpretation is possible, the one that

produces a closed figure will be preferred. This phenomenon is illustrated by the so-called Kanizsa diagrams shown in Fig. 3.15. These arrangements are usually perceived as a square (left) and a triangle (right), when in fact neither is actually present.

6. *Symmetry* Symmetrical areas tend to be perceived as objects against an asymmetrical background. This is illustrated in Fig. 3.16, where two black objects are usually perceived rather than three white objects.
7. *Law of Pragnanz* Given several possible interpretations of a scene, the best, simplest and most stable arrangement will be perceived (Koffka, 1935). The subjectivity of this law led to much criticism of Gestalt psychology. Just how are 'best', 'simplest' and 'most stable'

Figure 3.14 Illustration of the human preference for continuation.

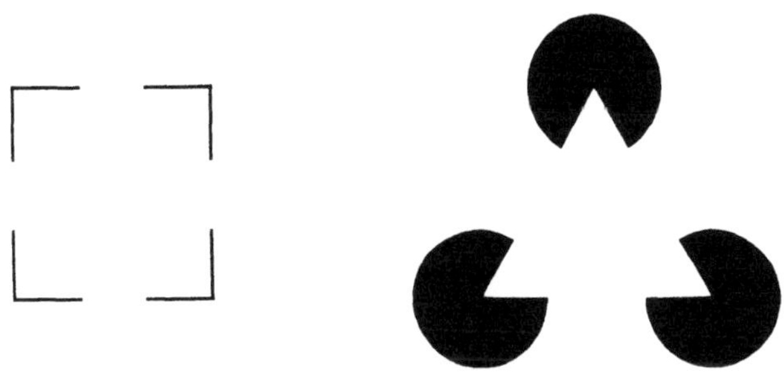

Figure 3.15 Illustration of the human preference for closure.

Figure 3.16 Illustration of the human preference for symmetry.

defined? Figure 3.17 is an attempt at illustrating this law in action. The top figure is usually perceived as three overlapping circular objects (the interpretation shown in the center figure) rather than one circular object and two circular objects with bites out of them (the interpretation shown in the bottom figure). The interpretation of this figure may also be explained, however, as a combination of continuation, similarity and symmetry.

More recent attempts have been made to quantify the Law of Pragnanz (Hochberg and Brooks, 1960). These attempts involved the study of the perception of drawings constructed of line segments, and they revealed a possible complexity measurement. The perceived complexity of a figure varied directly with the number of continuous line segments, the number of angles, and the number of different size angles present. Furthermore, simpler figures tended to be perceived as two-dimensional objects, whereas more complex figures were perceived as three-dimensional objects. Figure 3.18 illustrates this, where two views

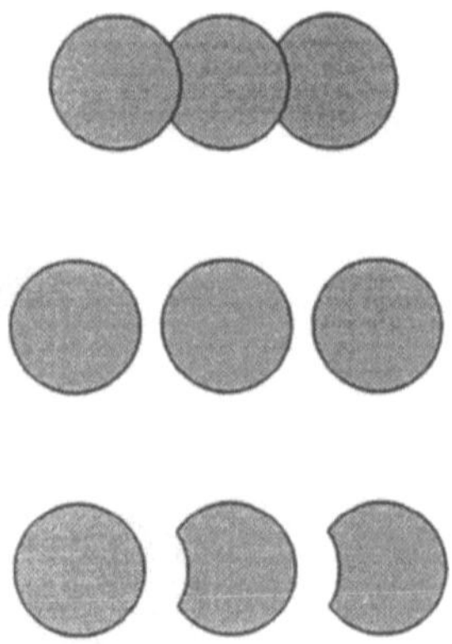

Figures 3.17 Law of Pragnanz illustration.

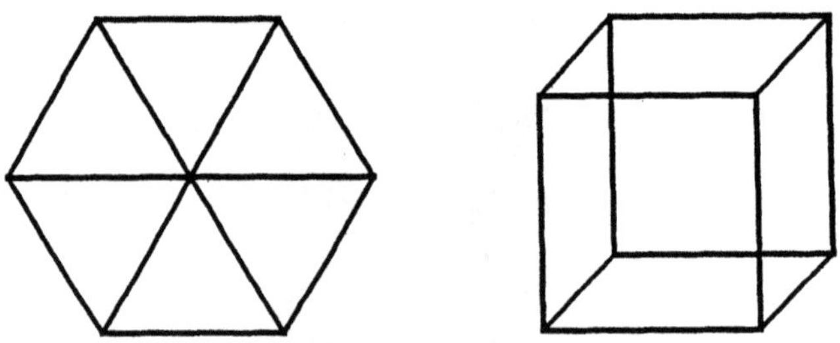

Figure 3.18 A wire-frame cube from two different perspectives.

of an identical wire-frame cube are presented. The view on the left tends to be perceived as a two-dimensional segmented hexagon, while the view on the right is generally perceived as a three-dimensional cube.

3.4 OBJECT RECOGNITION

3.4.1 Properties of objects

The question of how humans identify, or differentiate between, certain objects has led researchers to try and quantify the physical qualities that objects possess that make them differentiable. When a human being views a three-dimensional object, a two-dimensional array of wavelength and intensity is projected onto the retina. It is assumed, based on the previously presented physiological evidence, that an important part of early processing is the extraction of edges from this array. An edge is simply an abrupt change in intensity or color in the image. Edges are usually present around the perimeter of objects, but many other edges are detectable as well. Edges appear where there is an abrupt color change (as in the case of stripes on a zebra), or where an object has a distinct corner, or where a shadow is cast, to give a few of the possibilities. Edges divide the retinal array into regions, but what quality of these regions makes each of them unique and identifiable? The following list contains many suggested measurements of regions.

Number of line segments	Hochberg and McAlister, 1953
Number of angles	Hochberg and McAlister, 1953
Number of curves	Attneave, 1957
Symmetry	Attneave, 1957
Area (A)	Casperson, 1950
Perimeter (P)	Casperson, 1950
Maximum dimension	Casperson, 1950
Perimeter:area ratio (P:A)	Krauskopf *et al.*, 1954
Texture	Gibson, 1950
Dispersion (D)	Attneave and Arnoult, 1966

The dispersion measurement is defined by the following equation:

$$D = 1 - \frac{2\sqrt{\pi A}}{P} \tag{3.1}$$

Note that this quantity is independent of size, and that its value is zero for circles and approaches one for very irregular contours.

Each of the aforementioned measurements can be considered another dimension along which objects can be measured, in addition to the familiar Euclidean dimensions. Several researchers have suggested that there are 'psychological dimensions' along which the brain measures objects, and numerous studies have been conducted in order to determine the number and nature of these dimensions (Attneave, 1950; Attneave, 1957; Stilson, 1956). The idea is that the mammalian vision system performs some kind of transformation of an object from the more familiar physical dimensions into more useful psychological dimensions. One reason that the psychological dimensions may be more useful is that values for a specific object remain constant over a wide variety of viewing conditions, an attribute which many artificial vision systems seek to duplicate. This property is called **invariance**, and it is discussed in the following section.

3.4.2 Invariance

Human beings possess the ability to recognize a specific object in many different presentations. The image of the object on the retina may be large or small, it may be rotated by some arbitrary amount, and it may be in various positions, but the object is still recognized. The Gestalt psychologists called this phenomenon **isomorphism** (Koffka, 1935), and they argued that a specific object presents the same form to the brain regardless of position, orientation, or scale. Later psychologists called this phenomenon **stimulus equivalence** (Dodwell, 1971) or **shape constancy** (Palmer, 1983). All of these terms embody the same concept: that human beings can recognize objects irrespective of many changes in presentation. The four classes of invariance are described below:

1. *Position invariance* Objects are recognized regardless of their position on the retina, as illustrated in Fig. 3.19.
2. *Rotation invariance* Objects are recognized regardless of their rotation about some axis in three-dimensional space. A simple two-dimensional example is shown in Fig. 3.20.
3. *Scale invariance* Objects are recognized regardless of their relative size on the retina, as shown in Fig. 3.21.

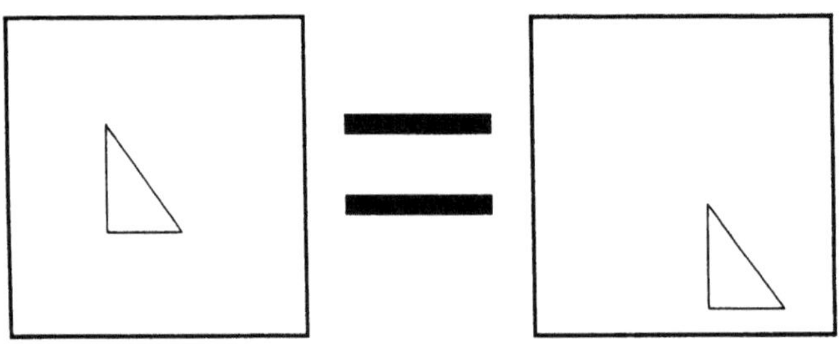

Figure 3.19 Objects that look similar regardless of position.

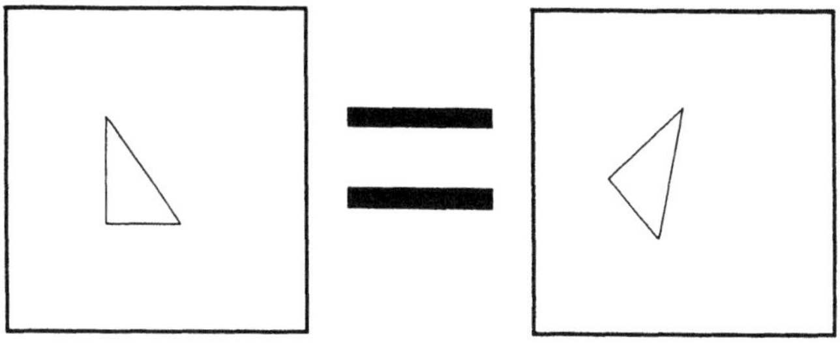

Figure 3.20 Objects that look similar regardless of rotation.

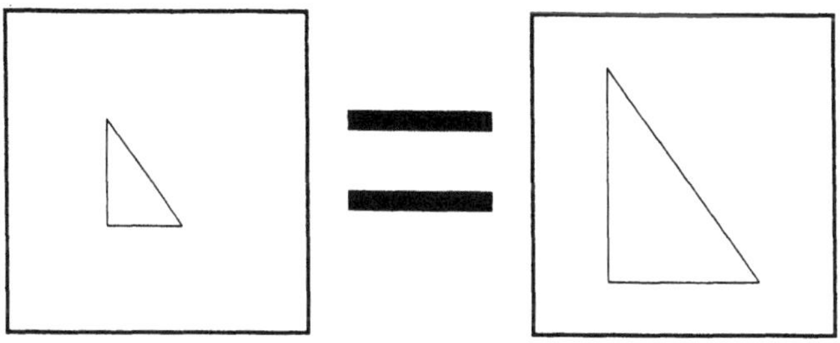

Figure 3.21 Objects that look similar regardless of scale.

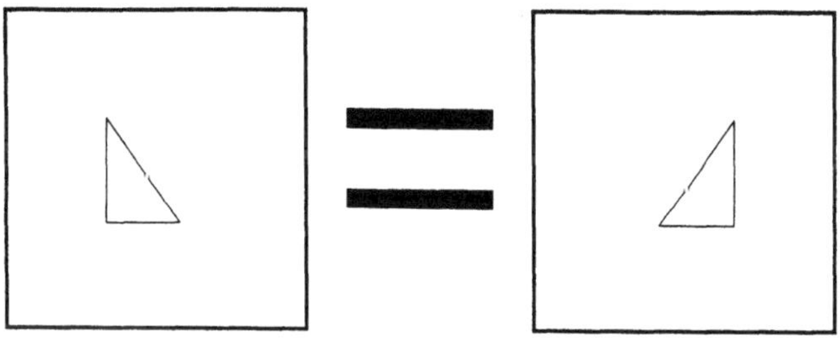

Figure 3.22 Mirror-image objects that look similar.

4. *Sense invariance* Objects are recognized even if the mirror image of the object is actually presented (Fig. 3.22). In fact, people often confuse objects with their mirror images (Duetsch, 1955).

Many vision researchers assume that object recognition in humans is totally invariant to all four of these aspects, but this is not so. Human perception is invariant to each of these aspects to varying degrees. Position invariance is good in the central visual area, but it falters somewhat in the peripheral regions (Duetsch, 1955). Rotation invariance is dependent upon the angle of rotation, and it seems to be based on experience (Mach, 1914). Size invariance is present only between extremes (Hake, 1966), and recognition of mirror images occurs less than half of the time (Dearborn, 1899).

Further experiments have shed some light on the possible mechanisms employed in providing invariance. Shepard amd Metzler (1971) conducted experiments in which subjects viewed pairs of images of novel three-dimensional objects (Fig. 3.23). The subject was to determine whether or not the images represented the same object in two different rotations. The time required to make this decision was approximately linear with the relative degree of rotation, which would suggest that some serial or repetitive process is required in order for a person to 'mentally rotate' one image to match another. These experiments suggested a mental rotation rate of approximately $60° s^{-1}$, and the results were almost identical for rotations in the image-plane and rotations in depth.

Similar experiments involving scale changes between images yielded similar results (Bundsen and Larsen, 1975). In those tests comparison time varied almost linearly with the degree of size disparity. As with the mental rotation process, these results suggest a serial or repetitive mental scaling process.

3.5 SUMMARY

The evolutionary process has developed a very impressive visual capability in human beings. A great deal of scientific effort has been directed towards discovering the structures employed in the human visual system, the function of those structures, and the techniques used to differentiate and identify objects in the two-dimensional visual field projected on the retina.

Objects can be represented by five attributes: position, rotation, scale, sense and shape (Palmer, 1983). Given some fixed frame of reference, the first four can be easily quantified. Shape, however, is most likely a

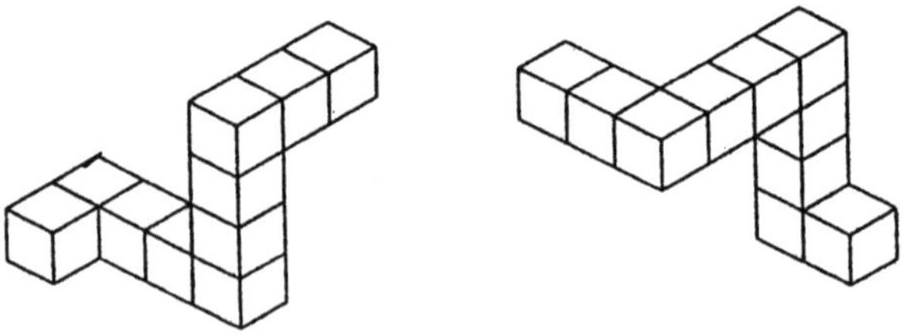

Figure 3.23 A pair of images similar to those used in mental rotation experiments.

multidimensional quantity based on some of the physical and psychological measurements presented in this chapter. Any artificial vision system that is to identify objects in a general environment, and therefore be useful as part of a general-purpose automated visual inspection system, will have to take all five of these attributes into account to some degree. Several systems that have attempted to do that are presented in Chapters 8 and 9.

4

Artificial neural networks for pattern recognition

4.1 INTRODUCTION

The predominant type of cells found in intelligent biological structures, like those responsible for vision in humans, are called **neurons** (Simpson, 1990). A biological neuron is illustrated in Fig. 4.1. Neurons are cells that consist of a cell body or **soma** that has appendages called **dendrites** that receive signals (inputs) from other neurons. The neuron cell has another appendage called the **axon**, along which the neuron sends its activation signal (output). The axon typically connects to the dendrites of many other neurons through junctions called **synapses**. Some of these synapses, when activated by the axon, tend to excite the receiving neuron into greater activity, while others tend to inhibit its activity. The human nervous system consists of approximately 10^{11} neurons, massively interconnected by a total of approximately 10^{14} synapses, with some connections extending over a meter in length (Wasserman, 1989).

Researchers have sought to duplicate human capabilities by creating models of these neurons and interconnecting them in various ways, giving birth to what are now known as artificial neural networks (ANNs). In this chapter the early history of ANNs is briefly described, and then several ANN paradigms that deal with the problems of pattern recognition and vision are reviewed.

4.2 EARLY ARTIFICIAL NEURAL NETWORKS

4.2.1 McCulloch–Pitts neuron

One of the first attempts at modeling biological neurons was conducted by McCulloch and Pitts (1945). The McCulloch–Pitts model treats

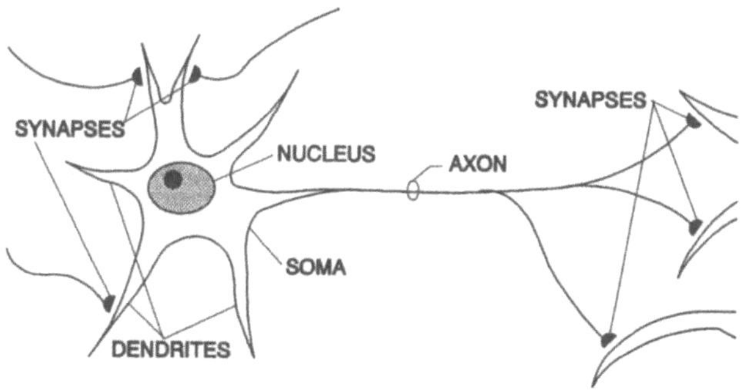

Figure 4.1 Structure of a biological neuron.

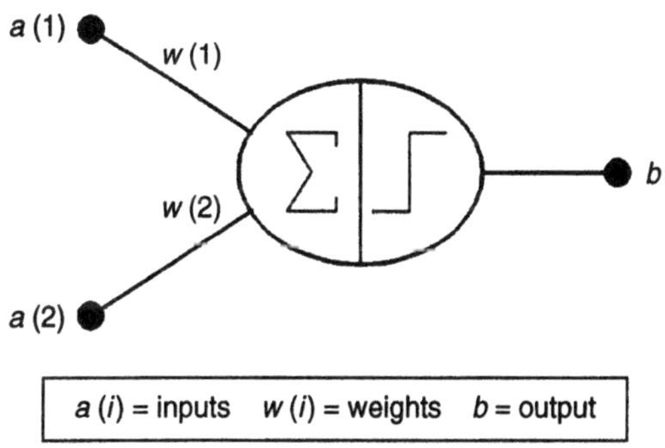

Figure 4.2 Structure of the McCulloch–Pitts neuron.

neurons as binary devices, that is they are either on (active) or off (inactive). In this model, neurons sum up their inputs, with each input being an activation from a previous neuron multiplied by some weight value representing the strength of the associated synapse. This sum-of-products is then passed through a threshold or step function, so that if the result is above the threshold the neuron's output value is one, and if it is below the threshold the output is zero. Figure 4.2 illustrates the basic McCulloch–Pitts neuron.

McCulloch and Pitts proved various theories about the behavior of these neurons, the most important of which was the fact that a network of these neurons could be constructed that would compute any possible Boolean function of its inputs. The proof treats neurons as two-input devices, and shows how, with proper weights and thresholds, neurons could be configured to behave like AND gates or OR gates. If inverted (inhibitory) inputs are allowed, networks of these two-input neurons can be shown to be capable of computing any Boolean function. This result is analogous to the well-known Boolean algebra proof that the operators (AND, OR and NOT) constitute a complete set (Rhyne, 1973).

McCulloch and Pitts later constructed networks of their neurons (Pitts and McCulloch, 1947) and investigated the properties of these networks. This represented the first systematic study of artificial neural networks. A network was even constructed for the purpose of pattern recognition, and a configuration was demonstrated that was capable of object recognition invariant to translation and rotation. This is the first known application of artificial neural networks to the problem of vision.

4.2.2 Perceptrons

Building on the work of McCulloch and Pitts, Rosenblatt (1958) and others (Widrow and Hoff, 1960; Widrow, 1961; Widrow and Angell, 1962) worked with single-layer artificial neural networks that became collectively known as **perceptrons**. Based on the McCulloch–Pitts neuron model, these networks consisted of a single layer of processing elements which had access to a common set of inputs, as illustrated in Fig. 4.3. The network in the figure appears to have two layers, but the lower (input) layer serves only to distribute the inputs to the processing layer, and conducts no processing of its own. Although many interesting properties of these networks were demonstrated, the most

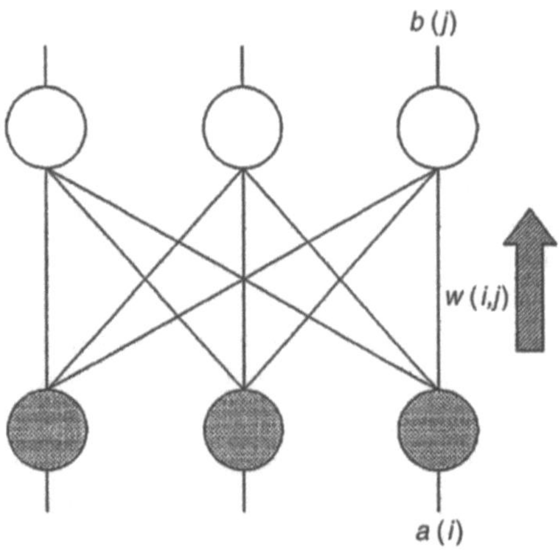

a (i) = inputs b (j) = outputs
w (i,j) = weight from input neuron i to output neuron j

Figure 4.3 Architecture of the perceptron neural network.

important contribution was the development of the perceptron learning rule (Rosenblatt, 1962). Using this rule, Rosenblatt proved that networks didn't have to be designed to perform a certain task, but that they could **learn** to perform any task that they were capable of performing. This exciting development greatly increased the research activity into ANNs.

As soon as research on perceptron neural networks began to flourish, however, researchers received some sobering news. In their landmark book *Perceptrons*, Minsky and Papert (1969) conducted a rigorous mathematical analysis of perceptrons, and proved that although they could learn any task that they were capable of performing, they were capable of performing only a very limited class of tasks. In particular, they proved that problems must be linearly separable in order to be solved by perceptrons. In the simple two-dimensional case (with two inputs to the network) this meant that perceptrons were only capable of dividing their input space (a plane) into two regions with a straight line. Problems that required other separations were out of reach. One such impossible problem is the exclusive – or (XOR) problem. Minsky and Papert showed that no combination of weights and thresholds could be found that would allow a perceptron to imitate a simple XOR gate. In high-dimensional cases (with greater than three inputs), perceptrons were shown to divide the input space only into regions separable by hyperplanes.

Although many interesting problems were linearly separable, many were proved not to be, and furthermore there was no easy way in which to determine if a problem fitted this criterion. This revelation put a serious damper on ANN research, and publications on the topic nearly vanished for over 10 years. Minsky and Papert also suggested that, although their work concerned only single-layer perceptrons, multilayer perceptrons probably suffered from similar limitations. Fortunately, this prediction later proved to be overly pessimistic. A hint of this could be seen in the earlier McCulloch and Pitts proof which showed that, although a single-layer perceptron may not be able to perform an XOR, a multiple layer connection of neurons was capable of performing **any** Boolean function.

4.3 BACKPROPAGATION NEURAL NETWORKS

4.3.1 Background

Although multilayer neural networks were known to be capable of many complex tasks that were beyond the reach of single-layer perceptrons, no rule existed which allowed multilayer networks to learn those tasks. One such learning rule, called **error backpropagation** or simply backpropagation, was discovered independently by at least three different researchers (Werbos, 1974; Parker, 1982; Rumelhart *et al.*, 1986). The discovery of this learning rule, and its convincing demonstration on many multilayer neural networks, was in part responsible for the resurgence of interest into ANN research.

4.3.2 Backpropagation network structure and processing

Backpropagation (BP) neural networks generally consist of three or more layers: an input layer, one or more hidden layers, and an output layer. The input layer usually performs a normalization function and distributes the inputs to the first hidden layer. Each layer is usually fully connected to the layer below it, that is each neuron in a layer receives inputs from all of the neurons in the previous layer. A typical three-layer BP neural network architecture is shown in Fig. 4.4. Although many variations of network structure and learning rules have been developed based on the BP concept, the standard or 'vanilla' backpropagation neural network will be explained here.

The input layer performs a normalizing function, converting all input vector values so they fall into the range (0, 1), and transmits the results via weighted connections to the hidden layer. The hidden neurons compute the sum of their weighted inputs, and this result is passed through a sigmoidal activation function (equation 4.1).

$$S(x) = \frac{1}{1 + e^{-x}} \tag{4.1}$$

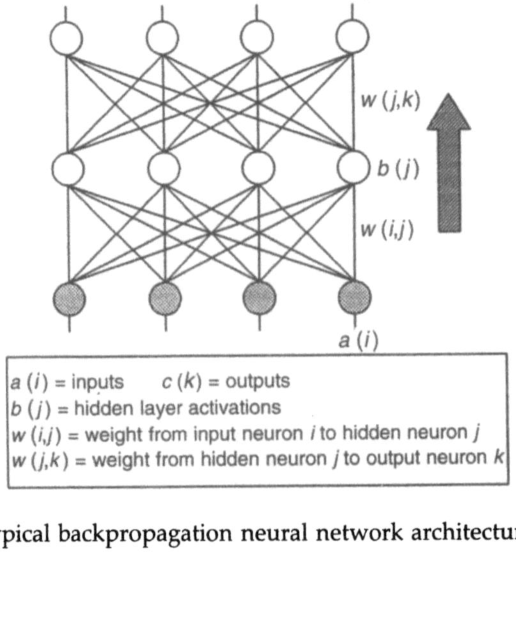

Figure 4.4 Typical backpropagation neural network architecture.

The processing done by each hidden neuron is shown in equation 4.2:

$$b_j = S(\Sigma_i N(a_i)w_{ij}) \tag{4.2}$$

where $N(x)$ is the input normalization function.

The sigmoidal function is shown graphically in Fig. 4.5. The fact that the sigmoidal function is differentiable is critical to the BP learning rule, as will be seen later. Note also that the sigmoid has a squashing or saturation effect, limiting neuron output values to the range (0, 1) regardless of input values.

The processing in the output layer is identical to that in the hidden layer, except the inputs come from the hidden layer and the weights used are those between the hidden and output layers. Equation 4.3 shows the processing performed by each output layer neuron. The outputs calculated in this manner become the network output vector in response to the applied input vector.

$$c_k = S(\Sigma_j b_j w_{jk}) \tag{4.3}$$

4.3.3 Backpropagation learning

Prior to the development of BP, the problem encountered when training a multilayer network was the proper assignment of blame for poor performance (or credit for good performance) to the hidden-layer neurons. When a network of this type is trained, a training set consisting of input–output vector pairs, with each input vector matched

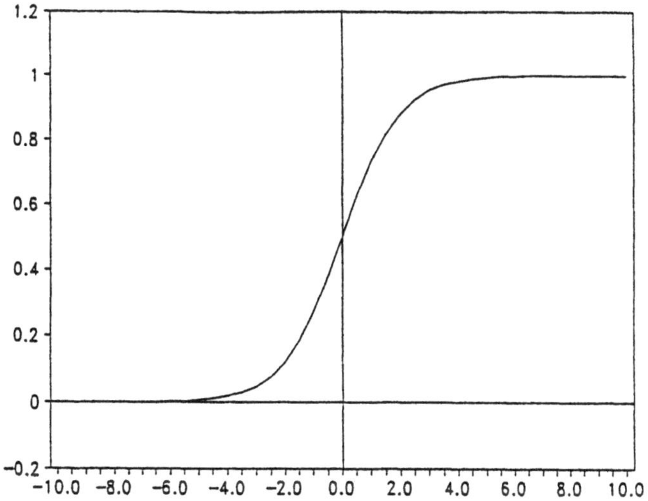

Figure 4.5 Sigmoid function used as a non-linear transfer function.

with the correct output vector, is usually used. An input vector is applied to the network, and the response of the network is calculated as previously described. If the response is not correct, the weights influencing the connections between the layers are adjusted by a learning rule to improve the performance. This process is repeated using each input–output pair in turn, and the complete training set may be presented several times. The process is continued until some error function representing the difference between the actual and expected outputs is minimized.

The assignment of error to the output layer neurons is simple: the derivative of the activation function times the difference between the actual value and the expected value is usually used. These error values are used to adjust the weights between the hidden and output layers. Equation 4.4 shows the error calculation used in the output layer:

$$\delta_k = c_k(1 - c_k)(c_k^* - c_k) \tag{4.4}$$

where c_k^* is the desired output at neuron k and δ_k is the error for output neuron k. Equation 4.5 shows how the weights between the hidden and output layers are updated:

$$\Delta w_{jk} = \alpha b_j \delta_k \tag{4.5}$$

where α is the learning rate.

The assignment of error to the hidden layer neurons is more difficult. The BP learning rule solves this problem by computing the partial derivative, with respect to each weighted input, of the output neuron's error value. The partial error values calculated in this manner are sent down to the hidden layer neurons. Each hidden layer neuron, then, receives error signals that are transmitted backwards from each output neuron. These error signals are then summed at each hidden layer neuron to determine the error value associated with that neuron, and the summed error values are used to adjust the weights between the input and hidden layers accordingly. If multiple hidden layers are used, this process is simply repeated at each hidden layer. Equation 4.6 shows how the error is calculated for each hidden layer neuron,

$$\delta_j = b_j(1 - b_j)\Sigma_k\, w_{jk}\delta_k \tag{4.6}$$

where δ_j is the error for hidden neuron j, and equation 4.7 shows how the weights between the input and hidden layers are updated.

$$\Delta w_{ij} = \alpha a_i \delta_j \tag{4.7}$$

In order to accomplish this type of learning, the activation function used in the neurons must be differentiable everywhere. This is one reason that the sigmoidal transfer function, rather than a threshold function, was necessary in BP neurons.

4.3.4 Summary of backpropagation

Backpropagation neural networks have been successfully applied to many problems. Among them are data compression (Cottrell *et al.*, 1987), painted surface inspection (Glover, 1988), speech synthesis (Sejnowski and Rosenberg, 1987), and character recognition (Burr, 1987). The popularity of BP stems from the fact that the mathematics of the network is well understood, and the general purpose nature of the learning rule allows its application to almost any problem. Another strength of BP is generalization. Because it is fully connected, the network learns to generalize a problem, and interpolates between trained pairs well. This ability also allows BP networks to act as good function approximators, and it is in this area that BP is presently applied most often.

Backpropagation is not without its difficulties, however. The BP learning rule is a member of a family of techniques known as **gradient descent**, in which each iteration of the procedure causes the network to progress 'downhill' on an error surface. If the error surface has many depressions distributed across it, and this is usually the case, the BP learning rule can become trapped in a local minimum rather than pursuing the global minimum error. Also, BP can require very long training times, since weight changes must be kept infinitesimally small to guarantee convergence of the learning process. Finally, the BP learning rule can lead to temporal instability in the network. This is a result of the fully connected nature of the network, whereby a change in any weight in the lower levels of the network affects the performance of all output neurons to some extent. For this reason, an input–output pair that previously performed well may perform poorly after additional training on unrelated vectors. Many neural networks have been designed that address one or more of the shortcomings inherent in backpropagation, and some of these are presented next.

4.4 SELF-ORGANIZING MAPS

4.4.1 Background

A great deal of organization has been experimentally observed in the human brain. Physical and mental functions can be mapped into specific areas of the brain, and these mappings are essentially constant across all human beings. Furthermore, similar functions are mapped to neighboring physical areas of the brain. A family of ANNs based on this concept, called self-organizing maps, has been developed by Kohonen (1984). These ANNs are capable of observing input patterns and organizing them, without outside assistance, into neighborhood groups internal to the network based on similarity found in the inputs.

4.4.2 Self-organizing map structure and processing

The basic self-organizing map (SOM) is a two-layer neural network (Simpson, 1990), as shown in Fig. 4.6. The horizontal connections in the output layer indicate the interactions between neighboring neurons. The input layer simply distributes the inputs, via weighted connections, to the output layer neurons. The output neurons compute the familiar sum-of-products calculation, summing the weighted inputs. A competition then takes place in the output layer which results in the neuron with the largest output winning, while all others are reduced to zero output. This is an example of what is called a **winner-take-all** competition (Freeman and Skapura, 1991). The winning output neuron, then, provides the only output from the network. The processing done in each output neuron is shown in equation 4.8, and the winner of the competition is determined as shown in equation 4.9:

$$b_j = \Sigma_i \, a_i w_{ij} \tag{4.8}$$

$$b_n = \begin{cases} 1 & \text{if } b_n = \max\{b_j\} \text{ A}j \\ 0 & \text{otherwise} \end{cases} \tag{4.9}$$

4.4.3 Self-organizing map learning

Learning in SOMs involves only the application of input vectors, not input–output pairs as was the case with the supervised learning employed by backpropagation. SOMs employ what is called **unsuper-**

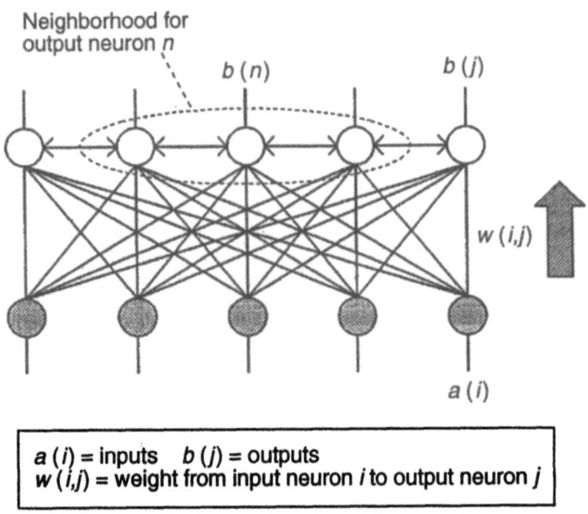

Figure 4.6 Self-organizing map neural network architecture.

vised learning, because outside influences other than the input vectors themselves are not required. In the first step of learning, an input vector is applied and the winning output neuron is determined as previously described. Once the winner has been determined, the neurons in its neighborhood are enabled by the lateral connections in the output layer. The extent of this neighborhood is determined before learning, and it is usually limited to the neurons immediately adjacent to the winner. Only the weights of the winning and neighborhood neurons, then, participate in the learning process. The weights are updated according to equation 4.10. The learning process is continued by repeatedly presenting input vectors to the network and conducting learning, with the learning rate α being reduced after each iteration according to equation 4.11.

$$\Delta w_{ij} = \alpha(t)[a_i - w_{ij}] \tag{4.10}$$

$$\alpha(t) = \frac{1}{t} \tag{4.11}$$

where t is the elapsed training time. This reduction in learning rate is necessary to guarantee the eventual convergence of the learning process.

The SOM learning technique has many advantages. First, only a limited number of weights are trained during each iteration, so learning already conducted on unrelated neurons is not affected. This leads to rapid training and temporal network stability. Second, because the neurons involved in any training step are physically in a certain neighborhood, the network assumes the structure of the inputs. This is called the topology-preserving effect, and it is the main reason that the network is said to be self-organizing. Without this effect (i.e. if only the winning neuron were trained), the network would randomly organize, with related inputs activating random output neurons.

4.4.4 Summary of self-organizing maps

The SOM paradigm has been applied successfully to many practical problems. A SOM has been used to recognize spoken words and translate them into written text (Kohonen, 1988). This system required as inputs for training only about 100 spoken words, and the network was trained in about 10 min on a personal computer. After the training phase, the network was able to transcribe speech into written text in real time with over 92% accuracy. Another application involved the learning of robot arm dynamics (Ritter *et al.*, 1989). The SOM in this application provided improved speed and robustness over traditional methods.

The SOM model is an example of unsupervised learning and self-organization in neural networks. These ideas are biologically plausible and have many practical applications, including pattern recognition and machine vision. The SOM also learns in such a way as to allow people to visualize what has been learned by observing the neighborhoods that have been formed. This is a major advantage over BP, which spreads learning evenly over all of the weights. Another ANN model that shares these advantages, and introduces even more, is presented in the following section.

4.5 ADAPTIVE RESONANCE THEORY AND ITS DERIVATIVES

4.5.1 Background

Human beings have the ability to remember something learned long ago, while still being able to learn new information. In contrast to this fact, most ANNs tend to forget or overwrite previously learned information while being trained on additional data. This problem is known as the **stability–plasticity dilemma** (Carpenter and Grossberg, 1987a). How can old memories be preserved in a stable manner, while our minds remain plastic so new ideas can be incorporated? This question represents one of the problems that was addressed with the formulation of the **adaptive resonance theory** (ART) (Carpenter and Grossberg, 1987a). As will be seen, ART combines the competitive learning strategy of self-organizing maps with a feedback and control mechanism that regulates memory coding. Weights are not changed unless the input pattern is sufficiently close to a pattern already stored by a particular output neuron. This pattern match is called **resonance**, and it is analogous to the resonance observed in mechanical systems at specific frequencies. By allowing learning only when resonance occurs, and by recruiting uncommitted neurons to code novel input patterns, ART solves the stability–plasticity dilemma.

4.5.2 ART-1 network

The first artificial neural network developed based on the adaptive resonance theory has become known as ART-1 (Carpenter and Grossberg, 1987a). ART-1 was designed for pattern recognition, and it was designed to deal with binary input patterns.

Network structure

The structure of a typical ART-1 neural network can be seen in Fig. 4.7. ART-1 works with binary input vectors, so inputs must be either 0 or 1.

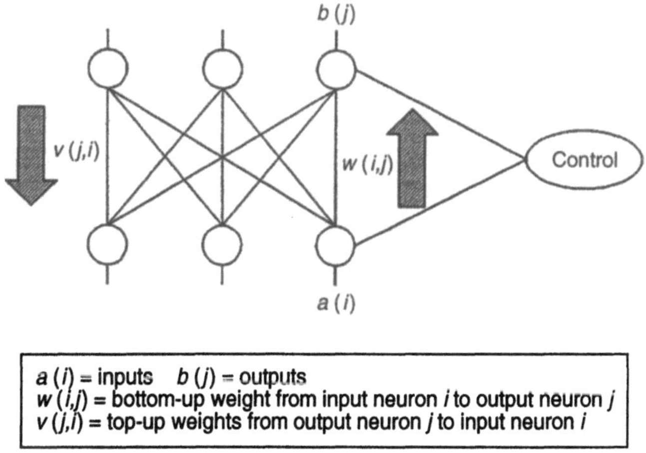

Figure 4.7 ART-1 neural network architecture.

Also, because ART-1 implements winner-take-all competition in the output layer, only one output neuron is active in response to a given input vector. The ART-1 network has two sets of weighted connections, one that is used in the usual bottom-up manner, and a second set that is used to create top-down feedback. ART-1 learning is conducted on-line, meaning that the network can modify its connections even while performing pattern recognition.

Recognition and learning

Forward or bottom-up processing in ART-1 is identical to that of the SOM described in the previous section, and it obeys equations 4.8 and 4.9. The recognition process is not complete, however, until the top-down processing and learning have been executed. Once a winning neuron has been selected in the output layer, the pattern stored in the top-down weights v_{ji} is fed back to the input layer. A comparison is then made between the input and feedback patterns. The fast learning version of ART-1 will be considered here, where the v_{ji} weights take on only binary values. In this case the comparison consists of a bitwise-AND between the two binary vectors:

$$x_i = a_i \wedge v_{ji} \; \forall i \tag{4.12}$$

To see if the match is good enough, the inequality below is tested:

$$\frac{|X|}{|A|} > \varrho \tag{4.13}$$

where $\varrho \leqslant 1.0$ is the vigilance parameter, $|X|$ is the number of 1 values in the bitwise-AND vector, and $|A|$ is the number of 1 values in the input vector. If the inequality is true, then resonance occurs, the pattern is recognized, and learning is conducted. If the inequality is not met, resonance does not occur, the winning neuron is removed from competition, and the competition in the output layer is conducted again. If no previously trained output node will resonate with the input, an uncommitted (untrained) node will be selected and trained. In this manner uncommitted nodes will be recruited to encode novel inputs, while previously stored patterns remain intact.

The vigilance parameter ϱ determines the degree of match required between the input and feedback (stored) patterns, and it ultimately determines the granularity of the classifications that the network will make. A low value of ϱ will tend to sort many input patterns into a few general categories, whereas setting $\varrho = 1.0$ will cause a separate category to be created for each non-identical input.

Once resonance has occurred, learning is allowed to proceed. Bottom-up weights associated with the winning output neuron are adjusted, as shown in equation 4.14:

$$w_{ij} = \begin{cases} \dfrac{L}{L - 1 + |X|} & \text{if } v_{ji} = 1 \text{ and } b_j = 1 \\ 0 & \text{otherwise} \end{cases} \tag{4.14}$$

where $L > 1$ is a learning constant. The top-down weights associated with the winning neuron are adjusted according to equation 4.15:

$$v_{ji} = \begin{cases} 1 \text{ if } x_i = 1 \\ 0 \text{ otherwise} \end{cases} \tag{4.15}$$

The learning process is conducted continuously while patterns are being recognized by the network. Because recognition and learning occur simultaneously, ART-1 is said to use on-line learning.

Summary of ART-1

ART-1 has been successfully applied to many image processing and pattern recognition tasks (Carpenter and Grossberg, 1988; Carpenter *et al.*, 1989; Rak and Kolodzy, 1988; Dagli and Vellanki, 1992). ART retains the advantages of self-organizing maps, while adding the ability to encode novel inputs without disturbing previously trained responses. The primary weakness of ART-1 is the fact that it is restricted to binary input patterns. ART has grown into a whole family of neural networks, however, with extensions being made to incorporate analog-valued inputs as well as many other enhancements. The properties of some of those ART derivatives are discussed in the following sections.

4.5.3 ART-2 network

The adaptive resonance theory has become a popular artificial neural network paradigm. Because of this wide acceptance, many enhancements and modifications have been made to ART so it could be applied to a wider domain of practical problems. One such enhancement was the adaptation of the ART neural network to enable it to handle analog-valued input vectors. The resulting network is called ART-2.

Network structure

The overall structure of an ART-2 network, developed by Carpenter and Grossberg (1987b), is identical to that of ART-1 (Fig. 4.7). The primary difference between the two networks is the fact that ART-2 was designed to handle analog-valued input vectors. This design change, although it may seem minor, adds considerable mathematical complexity to the network. Each of the input nodes is actually a supernode of sorts, containing several interneurons. These interneurons perform input and feedback vector normalizations and conduct the comparison between the input and feedback patterns. The internal structure of the input layer nodes is shown in Fig. 4.8.

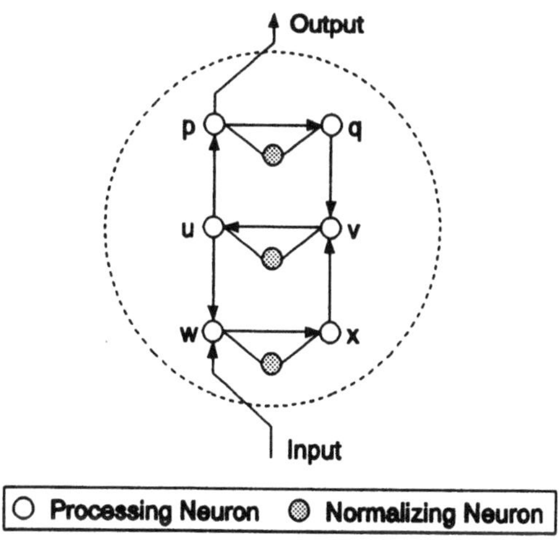

Figure 4.8 Internal structure of the ART-2 input layer neurons.

Recognition and learning

A brief qualitative description of ART-2 processing will be given next, but the level of mathematical complexity required to implement the network is much higher. During bottom-up processing, input vectors are normalized to unit length and each element of the vector is then passed through a threshold function. If the element is greater than the threshold it remains intact, but if it falls below the threshold it is set to zero. This thresholded vector is then passed through the bottom-up weights in the usual manner, and a winner-take-all competition is conducted in the output layer. The winning neuron feeds back a fixed signal through the top-down weights, generating the stored or reference vector. This vector is also normalized to unit length and then compared with the input vector. If the match is good enough (a vigilance parameter is again used) resonance occurs, recognition is indicated, and learning is conducted. If resonance does not occur, a search is conducted in a manner identical to that employed by ART-1. Learning involves rotating the bottom-up and top-down weight vectors associated with the winning neuron so they align more exactly with the input vector. This is done by changing each weight slightly in the direction of the corresponding element in the input vector. As with ART-1, learning is conducted continuously on-line.

Summary of ART-2

Like ART-1, ART-2 has been applied successfully to many practical problems in the areas of image processing and pattern recognition (Carpenter, Grossberg and Rosen, 1991a; Galindo and Michaux, 1990; Gjerdingen, 1990; Harvey *et al.*, 1990; Simpson, 1990). ART-2 preserves all of the advantages of ART-1 and handles analog-valued input vectors as well. One disadvantage of ART-2 is its considerable mathematical complexity, only hinted at in this simplified explanation. This level of complexity makes practical applications of ART-2 challenging.

4.5.4 Fuzzy ART network

In fuzzy ART (Carpenter, Grossberg and Rosen, 1991b), the principles of fuzzy set theory are applied to the ART-1 neural network. As in the case of ART-2, this modification allows the use of analog-valued input vectors. There are two primary differences between ART-1 and fuzzy ART. First, input vectors are analog-valued and are usually represented using **complement coding**. When complement coding is used, input vector elements are limited to values between 0 and 1. For each element a in the original input vector, the complement $a^c = 1 - a$ is calculated and concatenated onto the original vector. The resulting vector has twice as many elements as the original vector, and is automatically normalized.

The second difference is the fact that the bitwise-AND operations performed in ART-1 are replaced with fuzzy AND operations. The fuzzy AND conducts the MIN operation (Zadeh, 1965), where each element of the resulting vector is the minimum of the two corresponding elements of the vectors being combined. The effect of this fuzzy operation is that each output node represents a visualizable region in input space.

Network structure

As in the case of ART-2, the overall structure of a fuzzy ART network is identical to that of ART-1 (Fig. 4.7). The fact that the fuzzy ART usually employs complement coding, however, increases the size of the network. For problems involving input vectors of a certain size, the fuzzy ART network would have double the number of input elements as compared to the standard ART-1 network.

Recognition and learning

Bottom-up processing in fuzzy ART is identical to that of ART-1, except the fuzzy AND or minimum operation is used. The degree of match between an input pattern and the patterns stored at each output node is calculated using equation 4.16:

$$b_i = \frac{\Sigma_i \left|a_i \wedge w_{ij}\right|}{\beta + \Sigma_i \left|w_{ij}\right|} \tag{4.16}$$

where $\wedge$ is the fuzzy AND operation, and $\beta > 0$ is a choice parameter. The winner of the competition in the output layer is determined as follows:

$$b_n = \begin{cases} 1 \text{ if } b_n = \max\{b_j\} \ \forall j \\ 0 \text{ otherwise} \end{cases} \tag{4.17}$$

Again, the recognition process is not complete until the top-down processing and learning have been executed. Once a winning neuron has been selected in the output layer, the pattern stored in the weight vector that corresponds to the winning output neuron is fed back to the input layer. A comparison is then made between the input and feedback patterns. As with ART-1, the fast-learning version of fuzzy ART will be considered. In this case the comparison consists of a fuzzy AND between the weight vector corresponding to the winning output neuron and the input vector:

$$x_i = a_i \wedge w_{in} \ \forall i \tag{4.18}$$

The vigilance parameter is checked to see if the match is good enough by testing the following inequality:

$$\frac{\Sigma_i |x_i|}{\Sigma_i |a_i|} > \varrho \tag{4.19}$$

As with the other ART networks, if the inequality is true then resonance occurs, the pattern is recognized, and learning is conducted. If the inequality is not met, resonance does not occur, the winning neuron is removed from competition, and the competition in the output layer is conducted again. If no previously trained output node will resonate with the input, an uncommitted (untrained) node is selected and trained. Again, this guarantees that uncommitted nodes will be recruited to encode novel inputs and previously stored patterns will remain intact.

Once resonance has occurred, learning is allowed to proceed. The weight vector associated with the winning output neuron is adjusted as shown in equation 4.20.

$$w_{in} = x_i \tag{4.20}$$

As with other ART networks, the learning process is conducted on-line while patterns are being recognized by the network.

Summary of fuzzy ART

Fuzzy ART can handle analog inputs, and the introduction of complement coding creates automatically normalized input vectors, although in doing so it doubles the number of elements. Also, the introduction of fuzzy set theory concepts creates visualizable output regions. Finally, the changes introduced in fuzzy ART do not alter the stability/plasticity or learning convergence behavior that are the cornerstones of the adaptive resonance theory.

The development of the adaptive resonance theory is ongoing. Recent developments have included the addition of supervised learning to ART with ARTMAP (Carpenter, Grossberg and Reynolds, 1991), and the combination of fuzzy logic with supervised learning in fuzzy ARTMAP (Carpenter *et al.*, 1992).

4.6 NEOCOGNITRON NEURAL NETWORK

4.6.1 Background

Unlike ART and its derivatives, the neocognitron is a cross between a neural network and a model of the human vision system. The design of the neocognitron is based on the model of the mammalian visual system proposed by Hubel and Wiesel (1962), based on their research referred to in Chapter 3. The neocognitron was designed specifically as a pattern recognition device, so the inputs are two-dimensional images representing the retinal array.

4.6.2 Neocognitron architecture

The neocognitron is a multilayered architecture with a hierarchical organization, with layers lower in the hierarchy performing simple tasks and layers higher up performing more complex and abstract functions. A typical neocognitron architecture is shown in Fig. 4.9. Each layer in the neocognitron architecture is made up of a number of cells which are divided into two groups, simple cells and complex cells (Fukushima, 1980). Simple cells receive inputs from the previous layer, and the simple cells in the first layer receive inputs from the input layer or retina. Each simple cell is responsive to a certain feature in a specific position in the visual field. The simple cells are further divided into planes, with all of the cells in a given plane being responsive to the same feature but in different positions on the retina. All of the simple cells in a given plane usually have the same input weights.

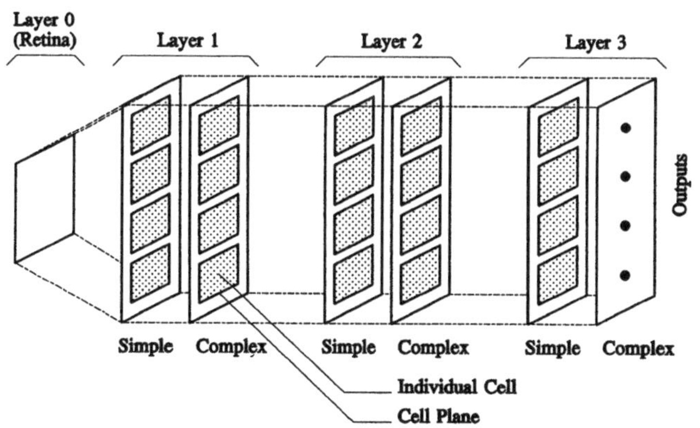

Figure 4.9 Neocognitron neural network architecture.

The complex cells in a given layer receive inputs from the simple cells of that layer. Complex cells effectively group the primitive features recognized by the simple cells, and therefore respond to more elaborate forms. Complex cells receive inputs from groups of simple cells in the current layer, and are therefore less position sensitive than the simple cells. Like the simple cells, complex cells are grouped into planes, with the cells in a given plane responding to similar objects. Complex cell input weights are usually fixed. As the layers become further removed from the retina, the complex cell response becomes increasingly abstract, to the point where a complex cell in the highest layer responds to a certain complex feature or pattern anywhere on the retina.

4.6.3 Neocognitron processing

The equations governing the processing done by the neocognitron are rather complicated and they will not be repeated here. Instead, a qualitative description of the processing done by each layer will be given. The exact mathematical details can be found in the original paper (Fukushima, 1980), or in the excellent coverage of the neocognitron by Freeman and Skapura (1991).

Simple cells respond to inputs in a given receptive region, usually a rectangular array. The output of a simple cell is calculated first by taking the weighted sum of the inputs in the usual manner. This result is then divided by the value of the input from an inhibitory cell, associated with the simple cell, which has the same receptive region. This inhibitory cell has fixed weights which are inversely proportional to the distance of each input from the center of the receptive region, and it provides a generic or average value to which the actual simple cell output can be compared for significance. The structure of the inhibitory cell weights also tends to make the simple cells more sensitive to features in the center of their receptive region. The result of this division is then passed through a linear threshold function, so the output of the simple cell is the calculated result if it is above the threshold, or is zero otherwise. This guarantees that only simple cells with excitations that are sufficiently above average have an output.

Complex cells also respond only to inputs within a given receptive region, and usually from a single simple-cell plane. Complex cell outputs are calculated in a manner almost identical to that employed by the simple cells, with the output of each complex cell being suppressed if it is not greater than a generic or average value. Later versions of the neocognitron have also incorporated lateral inhibition in the complex cell planes, where a cell with a high output suppresses the outputs of all of the other cells within its competition region, usually a rectangular or rhombic neighborhood centered on the winning cell (Freeman and

Skapura, 1991). In this way, the network can respond to several patterns simultaneously, as long as they can be recognized as different entities or they are physically separated.

4.6.4 Learning in the neocognitron

Several methods of learning have been employed with the neocognitron architecture, and both supervised and unsupervised techniques have been used (Wasserman, 1989). Usually, only the simple cell weights are trained, with complex cell weights remaining fixed. In the supervised case, each plane is preordained to respond to a certain feature, and when that feature is presented the cells on that layer are trained. Some form of Hebbian learning (Hebb, 1949) is usually used, with weight changes being proportional to the product of the input excitation and the output of the cell being trained. One cell in each layer is usually trained, with weights simply being copied to other cells in the same plane.

In the unsupervised case, the weights are initialized to small random values and the cell that responds most strongly to a given input is selected for training. A competitive learning technique similar to those employed by the SOMs and ART networks explained previously is used, where the weight vector is changed to more closely resemble the input vector during each training pass. Again, the weights for a single cell are trained and then copied to all of the cells in that plane. In this way the network self-organizes, and each plane becomes responsive to a different class of features.

4.6.5 Summary of neocognitron

The neocognitron has demonstrated impressive recognition capabilities. A four-layer neocognitron has successfully recognized handwritten numerals in the presence of positional shifts, noise and considerable distortion (Fukushima, 1988), and a compact one-layer network has demonstrated the ability to reliably recognize electronic components regardless of translational shifts (Dagli and Rosandich, 1993). Also, a large-scale self-organizing neocognitron network has been designed to recognize patterns regardless of scale, position and even a certain degree of overlap (Fukushima, Miyake and Ito, 1983). Although these accomplishments do not yet rival the performance of the human visual system, these results are impressive when compared to those demonstrated by other pattern recognition neural network paradigms.

Among the advantages of the neocognitron are its ability to recognize patterns, based on extracted features, invariant of scale and translation. The ability of the neocognitron to use either supervised or unsupervised learning based on data availability is also useful. The primary disadvantages of the neocognitron are its mathematical complexity and its immense size. A network with even a modestly sized retina can involve several thousand neurons, each with over a hundred weights (Fukushima, 1988). This fact makes practical applications of the neocognitron difficult to implement.

4.7 HAVNET NEURAL NETWORK

4.7.1 Introduction

This section presents a new neural network architecture, developed specifically for binary pattern recognition. Because the network utilizes the Hausdorff distance as a metric of similarity between patterns, and because it employs internally a version of the Voronoi surface to perform the comparison, it has been named the Hausdorff–Voronoi NETwork or HAVNET.

Similarity metrics

The choice of the Hausdorff distance as the metric of similarity between input patterns and learned patterns is what makes HAVNET different from most other neural network paradigms. Most current neural networks require two-dimensional input patterns to be converted into multidimensional vectors before training and recognition can be carried out. Learned patterns in these networks are represented as multidimensional vectors of trained weights, and the measure of similarity between the presented input and the learned pattern is based on some similarity metric. Three common similarity metrics in use in neural networks today are the Hamming distance, the vector dot product, and the Euclidean distance (Freeman and Skapura, 1991). In the case of binary {0,1} vectors, comparable measures of similarity between an input vector **A** and a weight vector **W** based on each of these metrics can be defined as follows:

$$\text{Hamming:} \quad 1 - \frac{d}{n} \tag{4.21}$$

$$\text{dot product:} \quad \frac{A \cdot W}{|A|} \tag{4.22}$$

$$\text{Euclidean:} \quad 1 - \sqrt{\frac{d}{n}} \tag{4.23}$$

where d is the number of mismatched elements between **A** and **W**, n is the number of total elements in **A**, and $|A|$ is the number of ones in vector **A**.

Unfortunately, transforming a two-dimensional input pattern into a multidimensional vector and then comparing that vector to learned vectors can produce behavior that is counterintuitive. Two-dimensional input patterns that appear very similar (to the human eye) to a particular learned pattern can generate very poor results when compared to that pattern using these metrics. Consider the pattern recognition problem illustrated in Fig. 4.10 in light of the previously defined similarity metrics. If the digit 'seven' shown in pattern (a) is considered to be the model, then the similarity of each of the other patterns to the model can be computed using the similarity metrics just defined. The results of these computations, as well as similar computations using a new metric based on the Hausdorff distance, can be seen in Table 4.1.

Note that, of all of the similarity metrics shown, only the Hausdorff metric exhibits a graceful reduction in value as the pattern deviates further from the ideal. This behavior is more representative of human performance than the behavior of the more common similarity metrics in use in current neural networks.

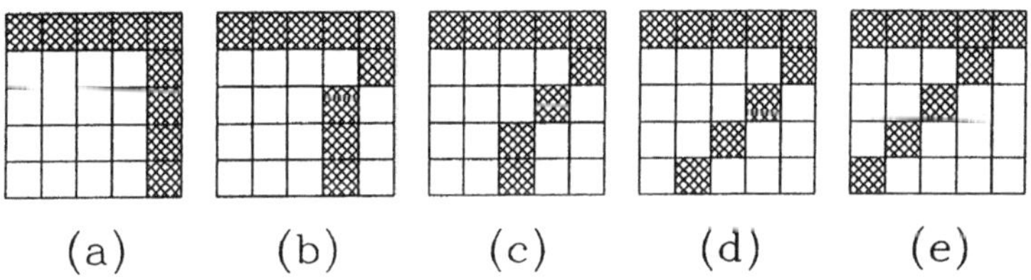

Figure 4.10 Different representations of the number seven for comparison.

Table 4.1 Values generated by different similarity metrics when comparing the various representations of the digit 'seven'

Similarity metric	*Pattern* (*a*)	(*b*)	(*c*)	(*d*)	(*e*)
Hamming	1.00	0.76	0.76	0.76	0.68
Dot product	1.00	0.67	0.67	0.67	0.56
Euclidean	1.00	0.51	0.51	0.51	0.43
Hausdorff	1.00	0.89	0.81	0.78	0.67

Hausdorff distance

The Hausdorff distance, when used as a measure of similarity between two-dimensional binary patterns, has been shown to agree closely with human performance (Huttenlocher, Klanderman and Rucklidge, 1993). The Hausdorff distance measures the extent to which each point of an input set lies near some point of a model set. Given two finite point sets $A = \{a_1, \ldots, a_p\}$ and $B = \{b_1, \ldots, b_q\}$, the Hausdorff distance is defined as

$$H(A, B) = \max\{h(A, B), h(B, A)\} \tag{4.24}$$

where the function $h(A, B)$ computes the **directed** Hausdorff distance from A to B as follows:

$$h(A, B) = \max_{a \in A}\left\{\min_{b \in B} \{\|a - b\|\}\right\} \tag{4.25}$$

where $\|a - b\|$ is typically the Euclidean distance between points a and b. The directed Hausdorff distance identifies that point in A that is furthest from any point in B and measures the distance from that point to its nearest neighbor in B. If $h(A, B) = d$, all points in A are within distance d of some point in B. The (undirected) Hausdorff distance, then, is the maximum of the two directed distances between two point sets A and B so that if the Hausdorff distance is d, then all points of set A are within distance d of some point in set B and vice versa.

The Hausdorff distance exhibits many desirable properties for pattern recognition. First, it is known to be a metric over the set of all closed, bounded sets (Csaszar, 1978). Also, it is everywhere non-negative and it obeys the properties of identity, symmetry and triangle inequality. In the context of pattern recognition this means that a shape is identical only to itself, that the order of comparison of two shapes does not matter, and that if two shapes are highly dissimilar they cannot both be similar to some third shape. This final property (triangle inequality) is particularly important for reliable pattern classification. It was because of these advantageous properties that the Hausdorff distance was chosen as the similarity metric that is the basis of HAVNET.

Three adaptations of the Hausdorff distance were required in order to make it appropriate for use in a neural network. First, the directed Hausdorff distance was computed on a point-wise basis. The directed Hausdorff distance was computed individually for each point in set A so that a single outlying point in A would not solely determine the result as it would in the setwise calculation. This adaptation increased the noise immunity of the measurement, a desirable property in neural network applications. The pointwise Hausdorff distance for a point $a \in A$ was computed as follows:

$$h(a, B) = \min_{b \in B} \{\|a - b\|\} \tag{4.26}$$

Second, each of the pointwise distances was truncated to limit the maximum distance and inverted so that a closer match generated a larger number. Truncation was done so that distances beyond a certain limit were ignored as insignificant, and the inversion was done so that close matches could generate higher activation levels in the neural network as is normally expected. The truncated and inverted pointwise Hausdorff distance is computed as follows:

$$h_\delta(a, B) = \begin{cases} \delta - h(a, B) & \text{if } h(a, B) < \delta \\ 0 & \text{otherwise} \end{cases} \tag{4.27}$$

where δ is the truncation distance.

Finally, the truncated and inverted pointwise distances were averaged for the point set A to compute the average as follows:

$$\overline{h_\delta}(A, B) = \frac{\sum_{a \in A} h_\delta(a, B)}{p} \tag{4.28}$$

where p is the number of points in set A.

Averaging has a normalizing effect on the measurement, making it invariant to the number of points present in the set. It also contributes to noise immunity and is partially responsible for the graceful reduction in value previously demonstrated. The similarity metric as defined in equation 4.28, the average truncated inverted pointwise directed Hausdorff distance, was used to compute the values previously shown in Table 4.1. More importantly, it has also been employed in the HAVNET neural network described next.

4.7.2 HAVNET architecture

An overview of the architecture of HAVNET is shown in Fig. 4.11. The neural network behaves as a binary pattern classifier. The HAVNET takes as inputs two-dimensional binary patterns, it employs feed-forward processing, and it produces analog output patterns. One output is generated by each node, with the analog value indicating the level of match between the input pattern and the class represented by that node.

The HAVNET neural network consists of three layers: the plastic layer, the Voronoi layer and the Hausdorff layer. The plastic layer contains several planes of neurons with the weights that are adjusted

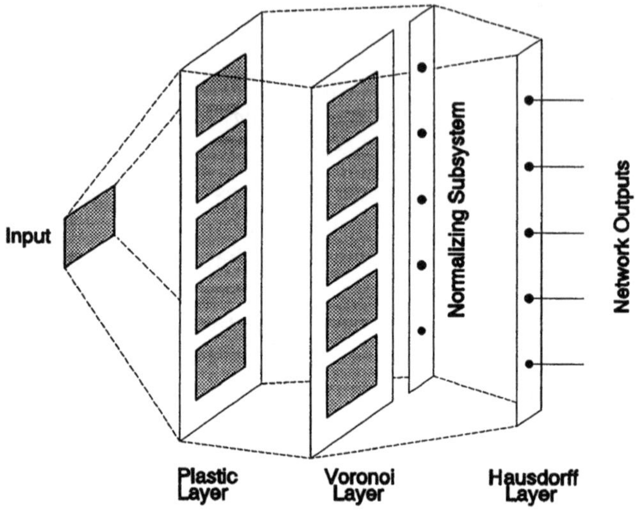

Figure 4.11 Overview of the HAVNET neural network architecture.

during the training process. One plastic layer plane is required to represent each pattern learned by the network. The Voronoi layer, utilizing a matrix of fixed weights, serves to measure the distance between individual points in the input and learned patterns. All of the planes in the Voronoi layer have identical fixed weights, but multiple planes are present to facilitate rapid parallel processing. The Hausdorff layer uses information generated by the Voronoi layer to compute the overall level of similarity between the input pattern and the learned patterns. One output node is present in the Hausdorff layer for each learned pattern. A normalizing subsystem is also present in the network, and it has the effect of making the response of the network invariant to the number of ones in the input pattern. It also negates the effect of uneven training between nodes, which can occur if a training set includes more examples of one category than another.

There are several similarities between the HAVNET architecture and that of the neocognitron (Fukushima, 1980; Fukushima, Miyake and Ito, 1983; Fukushima, 1988). The plastic layer is roughly analogous to a plane of simple cells in the neocognitron, as it receives input from the retina, contains multiple planes, and has trainable weights. The Voronoi layer is roughly analogous to a layer of complex neocognitron cells, since it has fixed weights and because each neuron in the Voronoi layer is associated with an area in the plastic layer, rather than a single neuron, which lends a generalizing effect to the network. Even the normalizing subsystem has an analog, as its behavior is similar to that of the V-cells in the neocognitron.

4.7.3 HAVNET learning

Learning is employed in HAVNET to adapt the individual nodes to recognize certain classes, and it is conducted by presenting examples of each class to the network during a training phase. Learning in this particular implementation of HAVNET is done off-line and in a supervised manner, but the network could employ on-line learning and self-organization if desired. During the learning process, a new node is added to the network whenever a novel pattern is encountered. The addition of a node involves the creation of a new plane in the plastic layer, a new plane in the Voronoi layer, and a new output node in the Hausdorff layer. The specific details of how learning and recognition are carried out are presented in the following paragraphs.

Off-line supervised learning was implemented in the HAVNET neural network because it was felt that this combination was best suited to most practical applications. The system is trained to recognize objects off-line, and the network is informed during training of the class to which each pattern belongs. Once the network is put into the recognition (on-line) mode, training ceases and recognition response is repeatable and predictable.

The weight matrices that undergo changes during the learning process reside in the plastic layer of the network. A binary input pattern A^m that is presented to the network during the learning phase is represented as follows:

$$a^m_{x,y} = (1, 0) \qquad x = 1 \ldots X; y = 1 \ldots Y \tag{4.29}$$

where X and Y are the dimensions of the input pattern. Prior to learning, each weight matrix w^n of the network is initialized as follows:

$$w^n_{(x+\delta),\,(y+\delta)} = 0 \qquad n = 1 \ldots N \tag{4.30}$$

$$w^n_0 = 0 \tag{4.31}$$

where N is the total number of nodes in the network. The quantity δ is defined as the span of the Voronoi layer, and it is identical to the truncation distance defined earlier in the discussion of the similarity metric employed by HAVNET. The truncation distance, as employed in the neural network, is a positive integer value that is somewhat less than the dimensions of the input pattern. The weight w^n_0 is defined as the averaging weight for a node, and it is trained during each training pass regardless of the input pattern. The w_0 weights serve as an indicator of the extent to which each node has received training, and they are employed by the normalizing subsystem.

When node n is trained on input pattern m, the change in each of the weights is computed as follows:

$$\Delta w^{n}_{(x+\delta),(y+\delta)} = a^{m}_{x,y}\,\alpha(1 - w^{n}_{(x+\delta),(y+\delta)}) \tag{4.32}$$

$$\Delta w^{n}_{0} = \alpha(1 - w^{n}_{0}) \tag{4.33}$$

where $0 \leqslant \alpha \leqslant 1$ is the learning rate.

Once the weight change is computed, the weights are updated as follows:

$$w^{n(t+1)}_{(x+\delta),(y+\delta)} = w^{n(t)}_{(x+\delta),(y+\delta)} + \Delta w^{n(t)}_{(x+\delta),(y+\delta)} \tag{4.34}$$

$$w^{n(t+1)}_{0} = w^{n(t)}_{0} + \Delta w^{n(t)}_{0} \tag{4.35}$$

where t is the number of training iterations. During the learning phase, each training pattern is presented to the network in sequence, and the appropriate node is trained using the equations above. The learning rate determines the magnitude of the effect that each training exemplar has on the trained weights.

In order to guarantee the stability of the learning process, it is necessary to prove that the training algorithm converges. Convergence is defined as a state in which, after a large number of training passes, the weights no longer change an appreciable amount, and furthermore that they converge on some constant value. The proof of convergence of the HAVNET learning process is given in Appendix B.

In addition to the learning just described, the HAVNET can employ competitive learning. Once the network has been trained on a number of exemplars, the trained weights can be adjusted using nearest-neighbor competition. This technique is similar to that employed in the self-organizing map family of neural networks (Kohonen, 1984). During the competitive learning phase, the excitation levels of the network output nodes that represent patterns other than the one presented for training are tested. If the node of this group with the highest output response is within a neighborhood parameter μ of the output generated by the correct node, the plastic layer weights corresponding to the incorrect node are reduced, thereby reducing the response of that node to future presentations of the current training pattern. In this way, the unique differences between nodes representing different patterns are emphasized and the similarities are downplayed.

Mathematically, competitive learning is carried out if the following inequality is satisfied:

$$net^c < net^i + \mu \tag{4.36}$$

where net^c is the output of the correct node, net^i is the output of the incorrect node, and μ is the neighborhood parameter.

Weight changes for the incorrect node are computed as follows:

$$\Delta w^{i}_{(x+\delta)(y+\delta)} = \alpha^{m}_{x,y}\alpha_c w^{i}_{(x+\delta)(y+\delta)} \tag{4.37}$$

where w^i are weights for the incorrect node, a^m is the current training pattern, and α_c is the competitive learning rate $(0 < \alpha_c < 1)$.

Weights for the incorrect node are then updated as follows:

$$w^{i(t+1)}_{(x+\delta)(y+\delta)} = w^{i(t)}_{(x+\delta)(y+\delta)} - \Delta w^{i(t)}_{(x+\delta)(y+\delta)} \tag{4.38}$$

where t indicates before competitive learning and $t + 1$ indicates after competitive learning.

In practice, competitive learning is carried out after some period of normal training, and it usually results in improved recognition performance of the network. Competitive learning enables the network to learn how to differentiate between similar patterns that represent different objects, for example between the digits 3 and 8. The competitive learning process can be continued until no further improvement in the recognition performance of the network is achieved. This entire training cycle can then be repeated if desired, with periods of competitive learning interleaved with periods of normal training.

4.7.4 HAVNET recognition

In the recognition mode, the HAVNET neural network attempts to classify an arbitrary input pattern into one of the classes represented by the trained network nodes. The network computes the previously defined modified version of the directed Hausdorff distance between an input pattern and a stored pattern at a given node.

To clarify the explanation of this computation, the concept of a truncated Voronoi surface is introduced. A Voronoi surface is constructed for a two-dimensional set of points A by first locating the members of A in the x–y plane, and then plotting in the third (z) dimension the distance from any point in the x–y plane to the nearest point that is a member of A (Huttenlocher, Kedem and Sharir, 1990). When this distance is not allowed to exceed some value δ, then the surface is defined as a truncated Voronoi surface. The plot of the truncated Voronoi surface is sometimes referred to as an egg-carton plot because, if the members of A form a rectangular grid, the resulting plot resembles an egg carton (Huttenlocher, Klanderman and Rucklidge, 1993).

The Voronoi surface can be used conveniently to compute the directed Hausdorff distance between two point sets. In order to perform the computation between a point set B and the previously defined point set A, the members of B are simply located in the x–y plane, and the z-value above each point represents the distance from that point in B to its nearest neighbor in A. The maximum of these values is the directed Hausdorff distance $h(B, A)$.

As previously mentioned, for neural network purposes it is desirable to compute the inverse of this distance, so that shorter distances result in higher outputs. It is also desirable to threshold this distance at some maximum, so that any distance above that maximum will generate the minimum network output (zero in this case). For these reasons it is desirable for the neural network to model the inverse of the truncated Voronoi surface. The truncation distance δ is the span of the Voronoi layer of the neural network that was previously referred to.

By internally generating this truncated and inverted Voronoi surface, the network is prepared to compute the truncated inverse pointwise directed Hausdorff distance (as defined in section 4.7.1) between an input pattern and a learned pattern, simply by projecting the learned

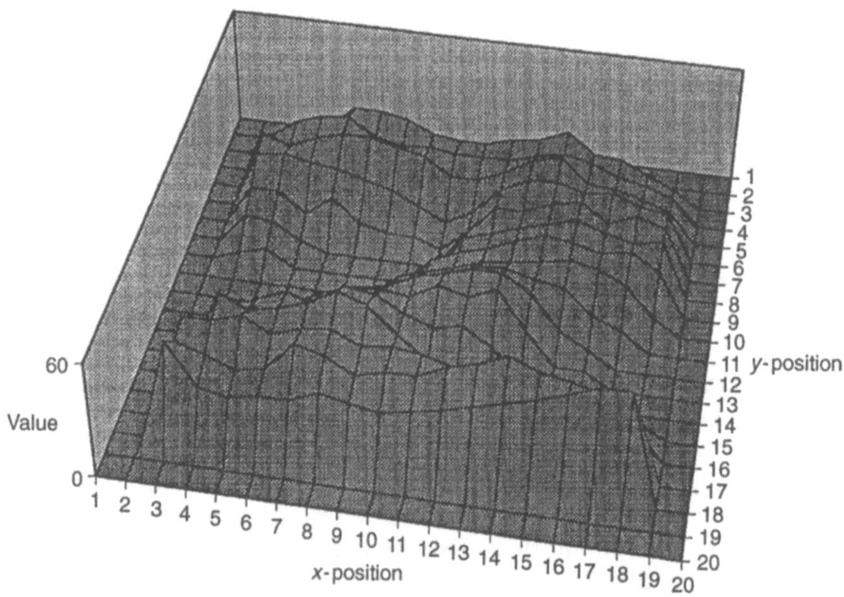

Figure 4.12 Plastic layer weights for a node trained to recognize the handwritten number two.

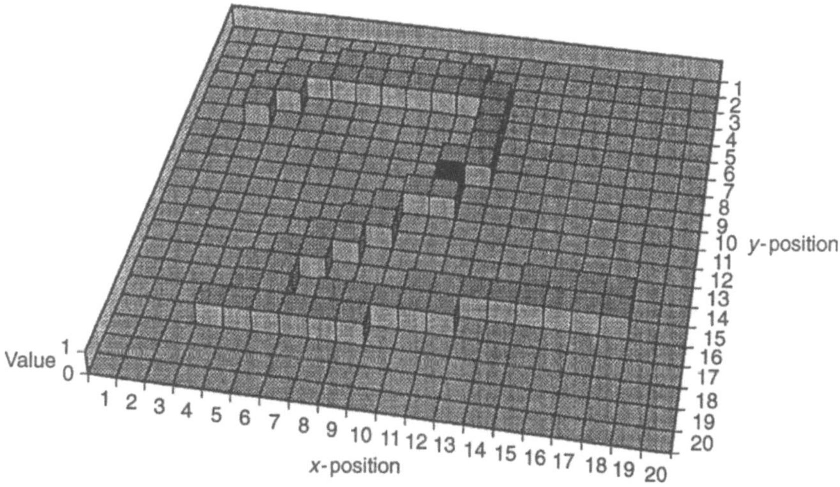

Figure 4.13 Example input pattern representing a handwritten number two, with a specific point in the input pattern highlighted.

weight values of the plastic layer onto this surface for each point in the input pattern. These pointwise distances are then averaged as previously described to compute the network output.

The recognition process can be illustrated graphically with a series of figures. First, consider a HAVNET node that has been trained on several examples of a handwritten number two. The resulting plastic layer weights are shown in Fig. 4.12. Then consider that it is desired to compute the response of that node to a new input pattern. The response is computed in a pointwise manner, so consider the computation for a single point in the input pattern. Figure 4.13 shows an input pattern with a single point highlighted. While computing the pointwise distance between this particular input pattern and the learned pattern, the network internally generates a truncated inverted Voronoi surface for the point of interest, as shown in Fig. 4.14. When the plastic layer weights are projected through this surface a response is generated as shown in Fig. 4.15. The maximum of that response surface represents the contribution of that particular point in the input pattern to the total calculation, which will be computed by averaging and normalizing similar responses over the entire input pattern. The mathematical details of this recognition process are given next.

The response of a node n to an input pattern A^m is determined by first computing the output of the plastic layer:

$$b^{n}_{(x+i),\,(y+j)} = w^{n}_{(x+i),\,(y+j)}\; a^{m}_{x,\,y} \tag{4.39}$$

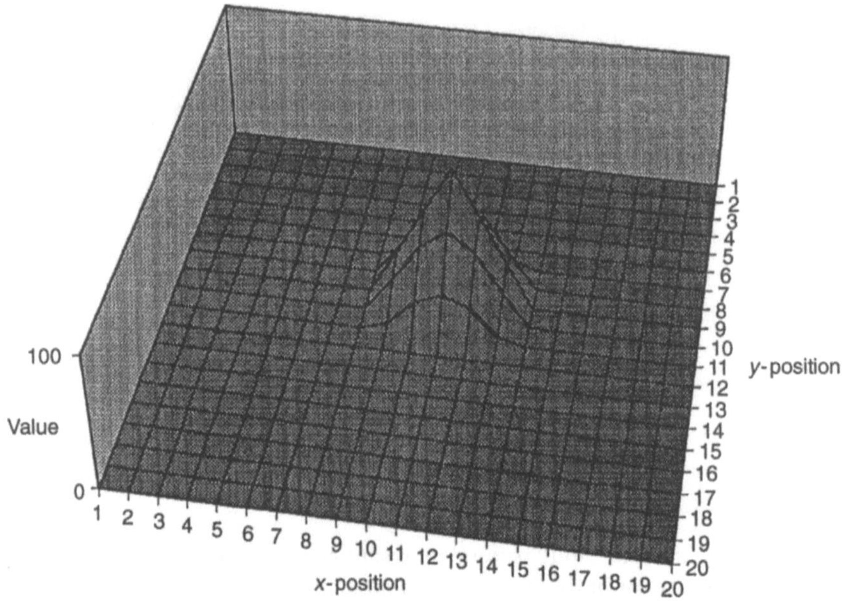

Figure 4.14 The truncated inverted Voronoi surface generated by the point of interest identified in Fig. 4.13.

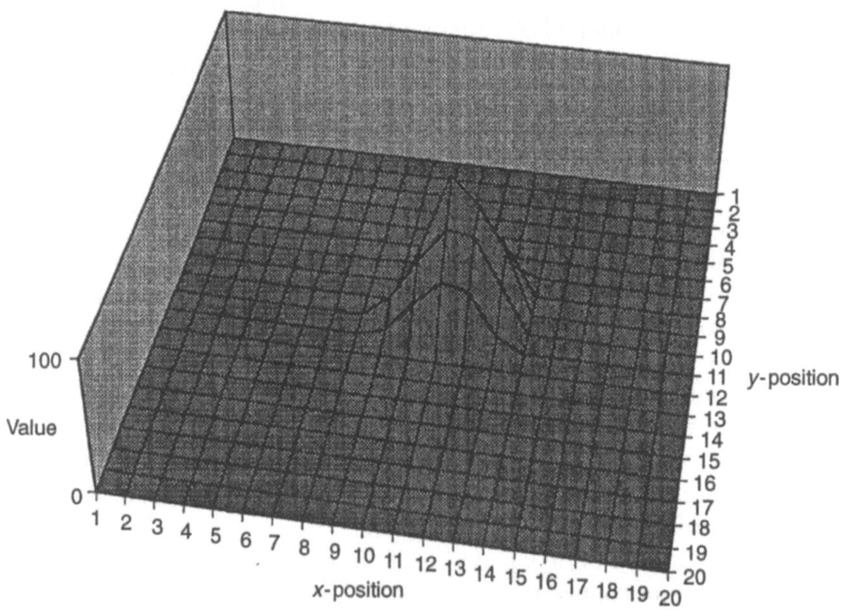

Figure 4.15 Response generated by projecting the plastic layer weights through the pointwise Voronoi surface.

where

x $=1 \ldots X$ is the input x dimension
y $=1 \ldots Y$ is the input y dimension
i, j $=-\delta \ldots \delta$ is the Voronoi layer span
n $=1 \ldots N$ is the node number
w^n are the plastic layer weights for node n
b^n are the plastic layer outputs for node n

Given the outputs from the plastic layer, the outputs for the Voronoi layer are computed as follows:

$$c^n_{x,y} = a^m_{x,y} max_j \left\{ \max_i \left\{ v_{i,j} b^n_{(x+i),(y+j)} \right\} \right\} \tag{4.40}$$

where c^n are the Voronoi layer outputs for node n.

The Voronoi weights $v_{i,j}$ are the same for all nodes and are computed as

$$v_{i,j} = 1 - \frac{\sqrt{i^2 + j^2}}{\delta + 1} - \delta \leqslant i, j \leqslant \delta \tag{4.41}$$

Once the outputs from the Voronoi layer are determined, the responses of the Hausdorff (or output) layer neurons are computed:

$$net^n = \frac{1}{w^n_0 p^m_a p^n_w} \sum_{y=1}^{Y} \sum_{x=1}^{X} c^n_{xy} \tag{4.42}$$

where w^n_0 is the averaging weight for node n, and net^n is the output for node n. Also,

$$p^m_a = \sum_{y=1}^{Y} \sum_{x=1}^{X} \phi(a_{x,y}) \tag{4.43}$$

$$p^n_w = \sum_{y=1}^{Y} \sum_{x=1}^{X} \phi(w^n_{(x+\delta),(y+\delta)}) \tag{4.44}$$

The function ϕ is the following binary threshold function:

$$\phi(x) = \begin{cases} 1 \text{ if } x > 0 \\ 0 \text{ otherwise} \end{cases} \tag{4.45}$$

The quantity outside of the summation in equation 4.42 represents the action of the normalizing subsystem and makes the node output invariant to the number of ones in the input vector and the past training history of the node. The final outputs net^n indicate, for each node, the similarity of the input pattern to the patterns that have been learned by that node.

4.7.5 Representing multiple aspects

The HAVNET neural network architecture just described has one major shortcoming: any class of objects is represented by a single network node. In many real-world applications members of a class of objects have several different appearances or guises, and it would be desirable to represent each of these in the network individually while still maintaining the overall identification of the class. One such application is character recognition, where different fonts may have quite different appearances. Another application is three-dimensional pattern recognition, where it is often desirable to represent several two-dimensional characteristic views of a three-dimensional object.

Such differing views or guises of objects of a class have been called aspects (Koenderick and van Doorn, 1979). In order to extend the HAVNET architecture to incorporate multiple views, the concept of aspects similar to that employed by Seibert and Waxman (1992) in visual object recognition has been utilized. The plastic layer of the HAVNET neural network has been expanded to include several aspect representations for each learned object rather than the single two-dimensional representation used previously, and the learning process has been modified to allow for the self-organization of these aspect representations. The recognition process of the expanded network has also been modified to accommodate the multiple-aspect object representation.

In order to incorporate learned knowledge of several aspects of an object, the extended HAVNET employs a plastic layer with the multidimensional structure shown in Fig. 4.16. The figure represents a matrix of aspect representations, with each aspect being represented by a two-dimensional matrix of trainable weights. The rows in the figure represent different object classes, and the columns represent the different aspects of each class. The number of learned aspects required to represent a class can vary depending on its complexity (all of the views of a featureless sphere, for example, could be represented by a single aspect). The number of aspects required is automatically determined by a self-organizing learning process.

Learning in the extended HAVNET involves an additional element to control the generation of new aspect representations in the plastic layer, a learning process patterned after that used in the ART family of neural networks (Carpenter and Grossberg, 1987a). Whenever a new view of an object is applied to the network for training, the outputs generated by each of the existing aspect representations for that object are tested. If the maximum of those is above a vigilance parameter ϱ, the match is

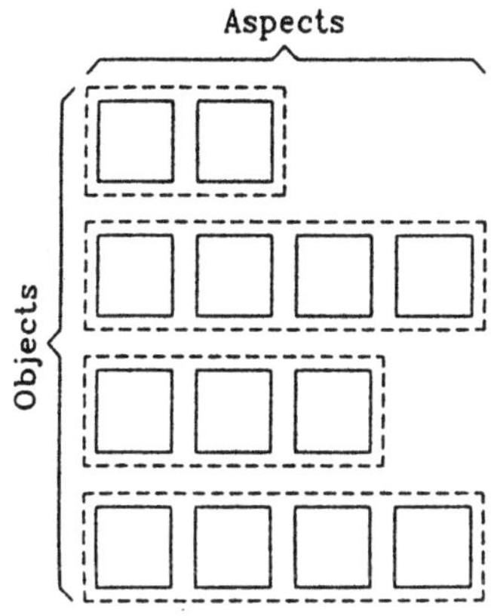

Figure 4.16 The plastic layer architecture of HAVNET extended to represent multiple aspects of learned objectes.

considered adequate and that aspect representation is trained on the input image. If the maximum output generated does not exceed the vigilance parameter, the match is considered inadequate and a new aspect representation is generated and trained on the input image.

The fact that a class of objects can be represented by multiple aspects in this extended version of HAVNET also leads to a slight modification of the recognition process. During recognition, the degrees of match between an input pattern and the various learned aspects of a class are calculated in the manner described previously, but the output of the node representing the class is computed based only on the aspect with the highest degree of match (i.e. the aspect generating the highest output). The output of the network as a whole, then, still represents the degree of match between the input pattern and each of the learned object classes.

4.7.6 A HAVNET example

In this section a simple example application of the HAVNET neural network is described. The objectives of this example are not to test or demonstrate the capabilities of HAVNET, but rather to better illustrate the learning and recognition processes and to clarify how the HAVNET can be applied to a practical pattern recognition task. Because of these objectives, the subject of this example will be a rather simple and familiar task, the recognition of digitized handwritten digits.

Inputs and outputs

A HAVNET network was designed to accept binary two-dimensional input patterns in a 16 × 16 format. The digitized digit images used in this demonstration were drawn from the AT&T Little 1200 database (Guyon *et al.*, 1992), because it was felt that these data are readily obtainable and therefore represent a reasonable benchmark task. The database consists of 1200 examples of individual digitized handwritten digits as sampled from 12 different writers. The database is divided into 10 groups of equal size labeled a–j, and each writer provided one example of each digit (0–9) for each group.

The HAVNET network was designed to classify the inputs into 10 categories, with one category representing each of the digits zero through nine. The output layer of the network consisted of 10 nodes, one representing each category. No pre-processing of any kind was performed on the input patterns prior to training or recognition. Some example input patterns are shown in Fig. 4.17.

Network implementation

The network used a Voronoi layer span of three, so that plastic layer weights a distance of three or more units from an input point would generate zero output. A unit is defined as one pixel in the input pattern. The Voronoi layer weights were fixed for all nodes and the weights used were as shown in Fig. 4.18. These weights are used by the network to compute the truncated inverted pointwise directed Hausdorff distance from an input pixel at the center of the pattern to any position in the plastic layer.

Because the span of the Voronoi layer was three, the effect of projecting the pointwise Voronoi surface onto the plastic layer extended two pixels beyond the 16 × 16 input size, so the plastic layer weight matrices were each sized at 20 × 20 to eliminate edge effects. This was done strictly for computational convenience, as no training took place in the two-pixel border area and the weights there remained at zero value. Whenever a new aspect representation was added to the network, another 20 × 20 plastic weight matrix was created. A normalizing weight was also created for each aspect to record the training history of that aspect. The normalizing weight was trained each time the aspect was trained, regardless of the configuration of the input pattern.

Because the automatic aspect generation technique used during training requires the network to perform a trial recognition on a training pattern, the recognition process will be described first. The degree of match between the input and any given aspect already represented in the network was computed using equations 4.39–4.45. Once all of the outputs were computed for the aspects representing a certain class, they were compared and the maximum was selected as the network output for that class. This procedure was repeated for each class represented in the network.

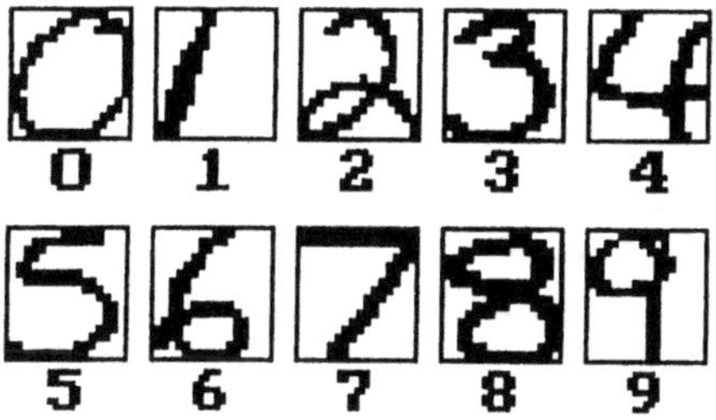

Figure 4.17 Typical input patterns representing digitized handwritten digits.

6	25	33	25	6
25	53	67	53	25
33	67	100	67	33
25	53	67	53	25
6	25	33	25	6

Figure 4.18 Voronoi weights used in the HAVNET example.

Supervised learning was used in this example, so the name of the class to which the input pattern belonged was presented to the network during training. As previously mentioned, before the network was trained on an input pattern a trial recognition was done in order to determine its level of similarity to the aspects already known to the network for the indicated class. If the maximum output exceeded the vigilance parameter, the existing aspect was trained on the new input pattern. If not, a new aspect was generated for the class and it was trained on the input pattern. In either case, the training was carried out using equations 4.32–4.35. When competitive learning was used to reduce the output of an aspect representing an incorrect class for a particular input pattern, it was carried out using equations 4.36–4.38. Before training was conducted all network weights were initialized to zero. The actual procedures used to conduct training and to test recognition on the selected data set are described in the following section.

Results and discussion

The database of handwritten digits was divided in half, with data groups a–e used for training and data groups f–j used for testing purposes. Initial learning of the inputs was accomplished by a single pass through the 600 input patterns in the training set, using a learning rate of 0.05 and a vigilance parameter of 60%. This initial training was followed by four passes of competitive learning with a learning rate of 0.10 and a neighborhood parameter of 0.0. In other words, if an aspect of an incorrect class generated a higher output than the highest aspect of the correct class, the incorrect aspect was trained competitively. The competitive learning was stopped after four passes through the training set because additional competitive training caused the recognition performance of the network on the training set to deteriorate. The final plastic layer weights for each aspect generated in the network during training are shown in Fig. 4.19.

After training, the recognition performance of the network was tested on the remaining data groups. The 600 remaining patterns were simply presented to the network, and the network outputs were recorded. The results are shown in Table 4.2.

Of the 600 test patterns, 545 were recognized correctly and 55 were recognized incorrectly for a recognition accuracy of 91%. For eight of the writers the number of errors out of 50 attempts was four or less, and

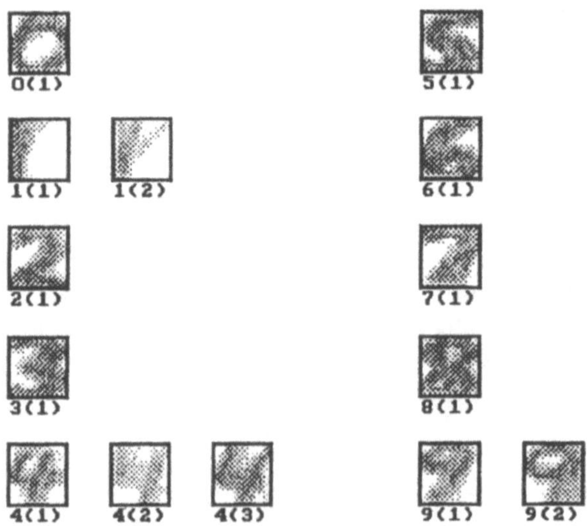

Figure 4.19 Plastic layer weights for the HAVNET example after initial and competitive training (the labels below the weight matrices indicate the node names, with the aspect numbers in parentheses).

the remaining four writers accounted for 39 of the 55 errors. The best recognition performance on a single writer was 100% and the worst was 70%. Figure 4.20 shows examples of some digits that were recognized correctly, and Fig. 4.21 shows examples of some that were incorrectly identified.

It should be noted that this network was not modified or optimized in any way for this particular task, and in fact it could be trained to recognize any input patterns of this particular size. Also, the recognition performance could probably have been improved with some input preprocessing, to reduce the effects of scaling and skewing.

Table 4.2 Recognition performance of the trained HAVNET network on the handwritten digit recognition task

Digit writer	*Number correct*	*Number incorrect*	*Success rate (%)*
Bill	47	3	94
Bob	41	9	82
Donnie	48	2	96
Greg	46	4	92
Hpg	49	1	98
Isa	48	2	96
Larry	48	2	96
Rich	42	8	84
Sara	43	7	86
Stuart	48	2	96
Tony	35	15	70
Wayne	50	0	100
Totals	545	55	91

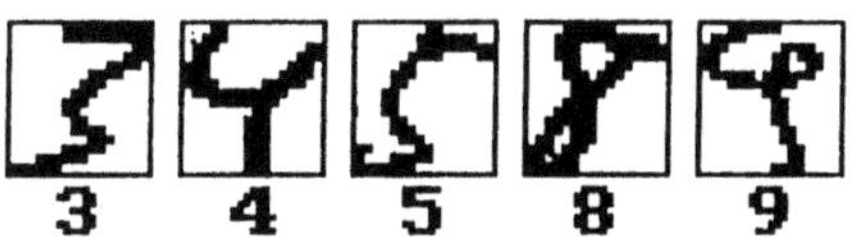

Figure 4.20 Examples of input patterns recognized correctly by the example HAVNET network.

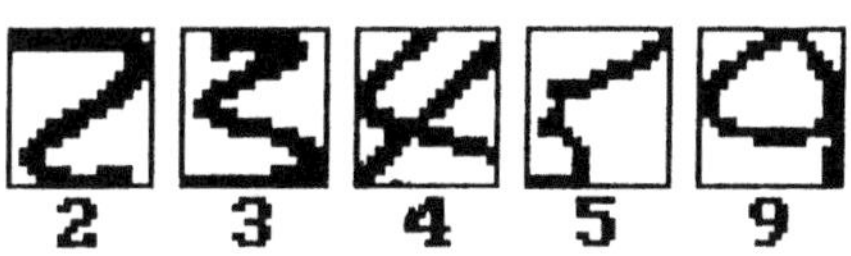

Figure 4.21 Examples of input patterns incorrectly identified by the example HAVNET network (labels indicate correct identification).

4.7.7 HAVNET summary

HAVNET is the first known neural network paradigm to take advantage of the Hausdorff distance as a metric of similarity between two-dimensional patterns. In doing so, the network inherits the desirable properties of the Hausdorff distance, and therefore duplicates human performance more accurately than most previous neural network architectures. The network is well developed and well behaved mathematically, and the network architecture is flexible enough to incorporate self-organization, unsupervised learning, and nearest-neighbor competitive learning in the plastic layer as required by specific applications. The network can also represent multiple aspects of a single object class, a feature that makes it much more useful in real-world object recognition applications.

Certain aspects of the HAVNET neural network are built upon past work in the neural network field. The architecture has much in common with that of the neocognitron, and the learning processes are similar to those employed in the ART and SOM networks. The use of aspects to represent characteristic views is also based on the approach used by Seibert and Waxman. In a more general sense, the cones generated by the Voronoi layer are similar to the fuzzy membership functions used in many applications, and they also resemble the radial basis functions that have recently been employed successfully in many neural networks.

4.8 SUMMARY

Pattern recognition is one of the most natural applications of artificial neural networks. Several artificial neural network models were presented in this chapter, each of which has demonstrated considerable capability in the pattern recognition domain. The HAVNET neural network was covered in greater depth because it is new and because it is used as the pattern recognition portion of an artificial vision system that will be detailed in later chapters.

In addition to the ability to recognize patterns, ANNs have demonstrated other capabilities that make them excellent choices for incorporation into artificial vision systems. For example, most ANNs exhibit a great deal of tolerance to imperfect inputs. Even with incomplete, noisy or inaccurate input data, ANNs can often produce good recognition results. The parallel processing utilized in ANNs also leads to a high level of robustness. Damage to a network in the form of malfunctioning neurons or corrupted weights usually results in what is known as **graceful degradation**, a loss of performance proportional to the loss of resources, as opposed to the catastrophic failure usually experienced in serial computer systems. Finally, ANNs can adapt to changing conditions by employing some form of learning. Learning could be used in automated inspection systems to compensate for things like changing levels of illumination, or to allow a system to learn the appearance of new defects so they could later be recognized. In short, the lack of flexibility, adaptability and robustness in current automated visual inspection systems could be addressed by developing systems based on artificial neural networks.

Part Three

Artificial Vision Systems Design

5

Image acquisition and storage

5.1 INTRODUCTION

As was explained in Chapter 3, the input for any vision-based sensing system is light intensity. The variation in light intensity over space or time provides information about the environment or product (Zuech and Miller, 1989). This chapter covers the various aspects of acquiring two-dimensional images and storing them for further processing. Although the hardware required to accomplish these tasks has become quite commonplace, the design of the image acquisition system should not be taken lightly. The quality and consistency of the acquired images have a dramatic impact on the level of difficulty that will be encountered in further processing, and many challenging image processing problems can be eliminated or at least minimized by careful detailed design of the image acquisition system.

The following sections present some of the details to be considered. Various camera types and lenses are covered, as well as methods of object illumination. Image acquisition cards are also discussed along with the image processing computers to which they are usually interfaced. Finally, image storage formats and compression techniques are explained.

5.2 CAMERAS

5.2.1 Vidicon cameras

The first video cameras were based on the Vidicon tube. In this type of camera, a two-dimensional image is focused on a small rectangular target of sensitive material by a lens. This target is then scanned by an electron beam, and the electrical signal from the target at any instant is proportional to the light intensity at the spot being scanned. The resulting signal can then be transmitted, processed or displayed on a cathode ray tube.

Although Vidicon-based cameras were very popular, they have some drawbacks. First, a point of high intensity in the image can saturate a large area of the target, obliterating the surrounding area of the image. Second, the tubes are temperature sensitive, since the geometry of the tube internals changes as it warms up, and the properties of the target material are somewhat temperature dependent. This fact causes drifting in image focus and variation in scale and aspect ratio, which cause great difficulties where an accurate image is required for inspection or measurement purposes. Vidicon tubes are also very delicate in construction, with extremely small components suspended in a glass envelope at very close tolerances. This makes them vibration sensitive, with vibration causing image blurring at best and tube destruction and camera failure at worst. Finally, like all tubes, a Vidicon is physically large, it requires a relatively high quantity of power to operate, and it generates a large amount of heat, making it cumbersome for portable applications.

5.2.2 Solid-state image sensors

Because the Vidicon tube has so many drawbacks, engineers have created solid-state alternatives for image sensing. There are primarily three technologies presently in use: the photodiode array, the charge-injection device (CID) array, and the charge-coupled device (CCD) array (Tseng, Ambrose and Fattahi, 1985). Because the CCD array is by far the most common sensor in use in industry today, it will be explained in more detail.

In a CCD array, light is sensed by an array of tiny photosensitive semiconductor devices. As light falls on the device, charge is accumulated in a way similar to the charging of a capacitor. The amount of charge accumulated by a specific device in a certain period of time is proportional to the amount of light falling on the device. In order to read the array, the accumulated charge at each device is transferred to a storage area where it is converted to a voltage. Because charge can be transferred between them, the light-sensitive device and the storage device are said to be **charge-coupled**, giving these sensors their name.

Two different techniques are used to read the photosensitive array in CCDs. In the first method, called **interline transfer**, each vertical column of devices transfers charge in parallel horizontally to storage areas located between the columns (Fig. 5.1). These storage areas then transfer their information vertically to a shift register, which transmits the video signal line by line. The advantage of interline transfer is that each light-sensitive device can spend most of its time accumulating charge because the transfer of charge to storage takes place in a very short period of time. This allows for excellent sensitivity to low-light conditions. The vertical shifting of information from the storage devices takes place while the next image frame is being formed in the light-sensitive devices. The disadvantage is that the storage areas must be fabricated between the columns of light-sensitive devices, so the density of the array is limited.

The other technique used to read a CCD array is called **frame transfer**. With this method the charge from each device is shifted serially downward very quickly into a storage area that it capable of storing the entire frame (Fig. 5.2). The information in the frame store is then transferred to a horizontal shift register that generates the video signal in a manner similar to that described above. The disadvantage of frame transfer devices is that the time required to transfer the frame of

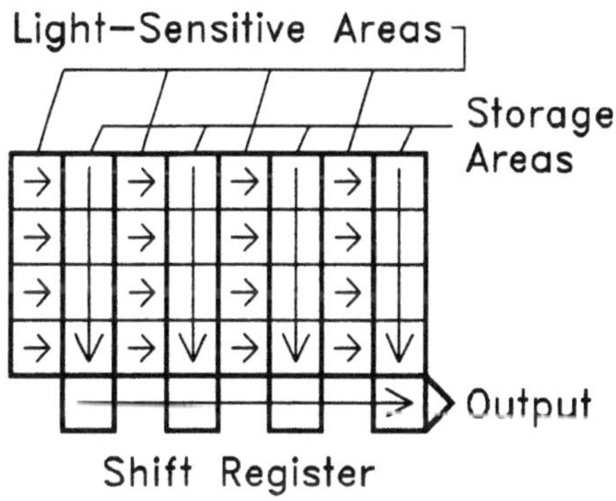

Figure 5.1 Interline transfer CCD array.

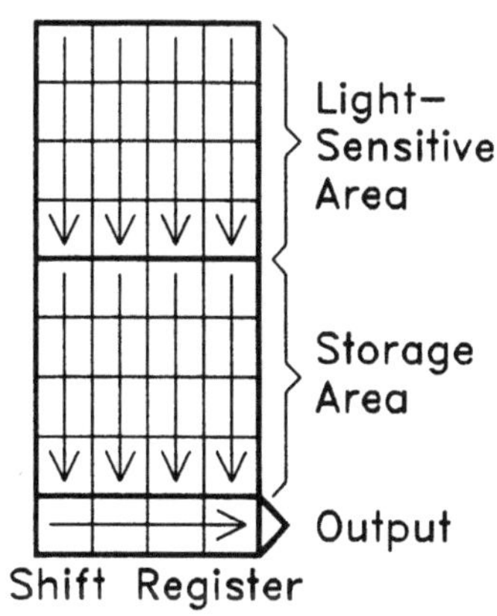

Figure 5.2 Frame transfer CCD array.

information to storage is greater, so less time is available for the light-sensitive device to accumulate charge. This leads to lower sensitivity to low-light conditions. The advantage is that no storage devices have to be fabricated in the light-sensitive array, so very high densities are possible. Cameras based on CCD arrays of each of these types are commercially available.

CCD arrays perform a function similar to that of the Vidicon tube, but they possess few of the drawbacks. Because CCDs are integrated solid-state devices, they are very small, require little power to operate, and generate very little heat. They are also extremely rugged, and operate over a wide temperature range. Once CCD arrays could be mass produced, low cost, rugged, and accurate solid-state video cameras became a reality. The development of inexpensive CCD arrays has resulted in the proliferation of home video cameras, and CCD array cameras are the preferred choice as input devices for industrial machine vision systems.

5.2.3 Line scan cameras

Because they have been used in many visual inspection systems, line scan cameras deserve at least brief mention. A line scan camera is a camera that uses a CCD array that consists of a single line of charge-coupled devices. This type of camera generates only a single line of image information, so either the object being imaged or the camera must be moving in order to generate a complete image. Line scan cameras are typically applied to situations where products are moving at constant speed, as on a conveyor belt. They are particularly good for making on-line width measurements of products being produced at a constant speed.

Line scan cameras have several advantages. They are simple in construction and therefore relatively inexpensive. The elimination of the complicated transfer and storage system present on two-dimensional CCD arrays allows the linear CCD arrays used in line scan cameras to be manufactured to very high resolution. The fact that they provide only a single line of image information per frame also allows them to operate at very high speeds. Line scan cameras also allow a continuous image to be created, as each line of image information can be stored indefinitely as a ribbon of image rather than a series of individual frames.

There are also disadvantages of line scan cameras. Because they do not generate a complete image, external equipment must be used to build up the image from multiple line scans. This image building process must also be synchronized with whatever motion is present in order to preserve scale information, and accurate measurements in the direction of motion are difficult. These cameras also transmit non-standard video signals that are incompatible with standard devices like monitors and video cassette recorders. Because of these drawbacks, and because cameras using two-dimensional CCD arrays have become more capable and less expensive, line scan cameras are currently losing popularity.

5.2.4 Digital cameras

Although CCD-based cameras are solid state, they are essentially analog in their operation. Charge values are stored as voltages which are transferred using analog shift registers to an amplifier that generates a standard analog video signal. Any digitization of the image takes place after it has been transmitted via the video signal. Because of this, there is no one-to-one correspondence between sensor elements on the CCD array and elements in the final image. For example, a camera with a 512 × 512 CCD array could be used to generate a 640 × 480 digitized image, and the precise source for the information in each element of the image would be uncertain.

Another alternative, the digital camera, is available. In these cameras, the signals from the CCD array are digitized immediately and transmitted as digital signals rather than analog video signals. Because of this, the light level at each position on the array is precisely represented by a digital number, and there is a one-to-one correspondence between the elements in the digital image generated and the devices on the CCD array, which leads to greater accuracy. The disadvantage is that the camera generates a unique digital signal for which there is no standard, so special interface hardware is required to interpret or convert the signal. As in the case of line scan cameras, the signal from a digital camera cannot be fed directly to a standard video monitor or VCR.

5.2.5 Camera lenses

Selecting a lens for a video camera is similar to selecting a lens for a still camera, so it will only be covered briefly. When selecting a lens for a specific camera, the mount type and lens type must be compatible. Most video cameras come with either a type C or a type CS lens mount configuration, and lenses of either type are readily available. Some cameras are available with a dual-purpose C/CS mount that can be converted to either configuration for maximum flexibility.

The best contact for information on how specific lenses will perform on a camera is the camera manufacturer. The manufacturers usually offer a selection of lenses that have been tested on their cameras, and are also aware of any idiosyncracies in combined camera–lens performance.

The primary specifications to consider when selecting a lens are focal length and light-gathering ability. The focal length is usually specified as a number of millimeters, and a camera will usually have a standard focal length. Lenses with focal lengths less than the standard are called **wide-angle** lenses and usually deliver a distorted 'fish-eye' image, whereas lenses with focal lengths longer than the standard are called **telephoto** lenses and are suitable for work at longer distances. Light-gathering ability is usually specified as an *f*-stop (aperture ratio) and is analogous to the same specification in still photography.

Special purpose magnifying lenses are also available for situations where extremely high resolution of detail is required. Magnifications of 30× are common using video cameras with high-magnification lenses. Cameras can also be adapted to many microscopes for imaging objects of even smaller scale. Magnifications of up to 1600× are common with camera–microscope combinations.

Many video camera lenses are equipped with automatic adjustments of some sort. The most common automated adjustments are iris, focus and zoom. Auto iris control allows the camera to compensate automatically for changing lighting conditions, and many manufacturers supply compatible lens–camera combinations that incorporate this feature. Auto focus allows a system to adapt automatically to different imaging distances. Motorized zoom lenses allow for automatic control of magnification, but these often require the automation of iris and focus as well since the adjustments are somewhat interdependent. Without these automatic adjustments the zoom, iris and focus of the lens must be set manually to match the conditions, and often must be readjusted if imaging conditions change.

5.3 OBJECT LIGHTING AND PRESENTATION

5.3.1 Illumination

Object illumination is one of the most important aspects of image acquisition. The type and level of illumination drastically affect the quality of the resulting image, and the problems of dealing with glare and shadows in images are difficult to overcome. It has been stated that selecting the proper type of illumination is the most important aspect of machine vision system design (Zuech and Miller, 1989). The simplest possible solution is to rely on ambient illumination. This seems plausible enough, since humans perform well over a wide range of lighting conditions, but machine vision systems have proved to have great difficulty with the varying illumination levels usually associated with ambient lighting.

The first variation on ambient illumination is some sort of controlled frontal illumination. With this approach, lighting of some sort is installed for the specific purpose of illuminating the scene being imaged, and the light source is placed behind or near the camera. The scene is then imaged using light reflected from the object. The level of illumination is usually relatively high in order to minimize the effects of ambient light, or the scene is somehow shaded or isolated from the ambient light. This system has the advantage of preserving the surface color and texture of an object, but it can suffer from unwanted glare and shadows. Because point sources of light, such as clear incandescent lamps, usually aggravate glare and generate sharp shadows, more diffuse light sources are usually used for frontal illumination. Frosted incandescent lamps, flood lamps and fluorescent tube lamps are all popular choices. An illumination ring, which is a circular light source that can be placed around a camera lens, is a particularly good choice for eliminating shadows from frontal illumination.

Another illumination variation often employed is backlighting. Backlighting involves placing a light source behind (from the camera's point of view) the object or objects being imaged, so the camera sees a bright background with the objects appearing as dark silhouettes. The use of backlighting often requires objects to be placed on a transparent or translucent surface illuminated from below. This system works best when the outline or profile of the object is important, and it eliminates shadows and glare effects. It also makes separation of the object from the background easy. All information about the surface of the object is lost, however, and transparent or translucent objects cause obvious problems. The transparent or translucent surfaces required for backlighting also have many practical problems in production environments. They must be kept clean, they are often fragile and therefore prone to breaking, and they can scratch easily. Finally, some modification of the normal product transport system is inevitable when introducing backlighting.

When sources of light need to be precisely controlled, fiberoptic illumination can be used (Hollingum, 1984). With this technique light is transmitted by optical fibers toward the object, and multiple sources can be used and directed where they are most effective. This system can provide excellent illumination with no shadow or glare problems, and can still reveal surface features and color. Previous information about the object or objects to be imaged is required, however, and a unique system is usually required for each type of object. This limits fiberoptic applications to very controlled cases.

Structured light is often used when three-dimensional properties of an object must be imaged. A curtain of light can be swept over an object (or the object passed underneath it) to generate a profile of an object from multiple images (Marshall and Martin, 1992). Grids of light can also be projected onto objects in order to reveal their three-dimensional shape. Fiberoptics or lasers are usually used to generate the desired structured light pattern.

5.3.2 Object presentation

Objects can be presented to a vision system in different ways in order to provide information in one, two or three dimensions. One-dimensional systems usually involve a linear strip source of light placed behind the camera, or the use of a line scan camera. The camera measures the amount of the source that is obscured by the object. This type of system is usually used to measure the thickness or width of a product being manufactured.

A two-dimensional presentation is more common. With this sytem a camera is simply pointed at the objects to be imaged, and the objects are illuminated using one of the methods previously described. The majority of the machine vision systems in use today employ some variation of the two-dimensional approach.

When depth information is needed, a three-dimensional presentation is required. This can be done with a single camera if the previously described structured light system is used, or with multiple cameras. Multiple images from cameras at different viewpoints can be compared to determine an object's three-dimensional shape or its exact location. An example of this is a two-camera stereo vision system that simulates human vision, a situation that will be discussed further in Chapter 7.

In order to acquire a usable image, a camera must be in the proper relationship with the object of interest. To accomplish this, a camera can be positioned so that objects pass through its field of view, or objects can be presented to the camera by a robot manipulator. Conversely, the camera can be mounted on a robot arm so it can be moved with respect to the object, as required.

In virtually every case where automated inspection is to be installed in a manufacturing environment, some modification of the product transport system is required for proper presentation of the product. Often the background must be cleaned up in order to eliminate unwanted or distracting objects from the field of view. Products that are in contact with each other or overlapping must often be separated so that they can be reliably imaged, and fast moving products must sometimes be momentarily stopped so that clear images can be obtained. Often shading or screening must be used to eliminate unwanted ambient lighting effects, and this can obscure product flow from the view of machine operators. Any modification of normal product flow has the potential to create a new plugging point in the product flow stream, also not a popular development as far as machine operators are concerned. When approaching the design of an automated inspection system, it is best to address realistically the changes in product handling or flow that may be required, and to do it at an early stage in system development. Any proposed changes should be communicated to management, production, and operations personnel as soon as possible in order to solicit comments or feedback and to facilitate acceptance of the system.

5.4 IMAGE ACQUISITION CARDS

5.4.1 Background

The video signal coming from a camera must be converted into a digitized image, and this is most commonly performed by an image acquisition card. This is a card that provides an interface between image processing hardware (usually some kind of computer) and the real-world video signal. This card can often be installed in the computer, and it is sometimes called a **frame grabber** because it captures individual image frames from the continuous video signal. Frame grabbers are available for the popular personal computer (PC) busses (ISA, Macintosh, Micro Channel, VESA and PCI). They are also available for industry standard busses such as VME and for many popular workstations. Continuing dramatic improvements in the performance of PCs have made them inexpensive and powerful computers capable of many image processing tasks, so the emphasis in this chapter will be on products intended for use in personal computers.

5.4.2 Operation

A block diagram for a typical frame grabber is shown in Fig. 5.3. The following description of frame grabber operation is based on that diagram and is representative of a large number of products available commercially. Any specific product, however, is sure to deviate from this description to some degree, and the detailed features of a board should be well understood before specification or purchasing decisions are made.

The incoming video signal is first passed through a high speed analog-to-digital converter so the continuous signal can be digitized. Converters with 8-bit or 12-bit outputs are common. This conversion process must be synchronized so that digitization begins at the beginning of a frame (not in the middle) and so that individual lines of image information are preserved correctly. Synchronization can be provided internally on the board, in which case the incoming video signal is analyzed for a pattern that indicates the beginning of a new frame as a signal to start digitization, and each line is digitized until an end-of-line pattern is

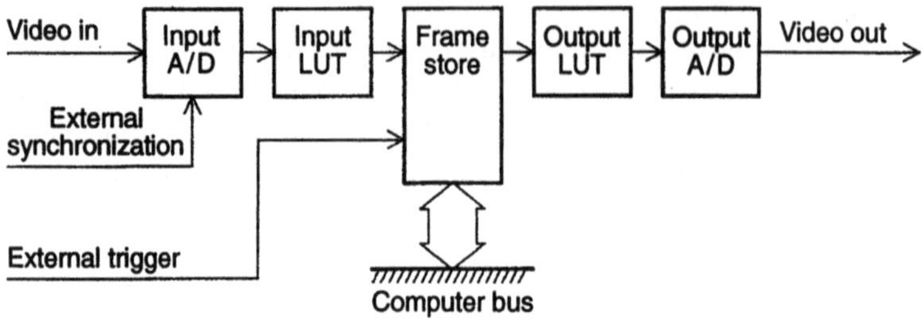

Figure 5.3 Block diagram for a typical frame grabber board.

detected. External synchronization can also be used, where an external signal indicates when frames and lines begin. This is often used when both the camera and the frame grabber must be synchronized to some external event.

The digitized signal is then passed through an input look-up table (LUT). An 8-bit signal, for example, would have values 0–255, and these are used as indexes into the LUT. The value stored in the LUT at each index is the actual pixel value that is sent to the frame store. If the image is to be simply passed through in its original form, the values in the LUT will equal the index numbers. Other LUT values can be used, however, to perform simple image pre-processing at this stage.

The values from the input LUT are sent to the frame store, which is a section of memory large enough to store at least one complete image. The frame store can often be used to store multiple images if enough memory is present. The frame store can be operated in two modes, continuous and triggered. In the continuous storage mode, each incoming frame is stored as it is digitized, replacing the previous frame in the store. In the triggered mode a frame is stored only after a trigger signal is received, and all other frames are ignored. Triggered storage is usually used in inspection systems, since some external event like the arrival of a product is usually used to trigger the system to capture and store a snapshot of the scene. This stored frame can then be transferred to the computer and analyzed, and it is not overwritten until another trigger signal is received. The frame store memory is usually connected in parallel to the computer bus so that complete frames can be transferred from the frame grabber board to the image processing computer's memory quickly. Data from the frame store is also passed through an output LUT where a look-up process similar to that employed in the input LUT is used to perform post-processing of the image prior to display. For example, pixels of a certain intensity range could be highlighted by changing them to a certain color in the output LUT. Data from the output LUT is then passed through a digital-to-analog converter to recreate the video signal for display on a monitor. In this way the captured images stored in the frame store can be viewed in real time, with the benefit of any post-processing performed in the output LUT.

5.4.3 Choosing an image acquisition card

The first thing to consider when choosing an image acquisition card is the bus standard that it is designed to interface with. The board must be compatible with the computer that will be used to store and process the images, and some common bus standards have already been mentioned. Within any particular standard there are also variations that must be considered. The number of bits used, the power available, the size and number of slots available, and the bus speed are all variable to some degree. Cards installed in bus architectures usually interface with the host computer via interrupts, and the availability and suitability of interrupt lines in the host system must also be considered. Some bus architectures also require input–output (I/O) address space to be allocated to each board on the bus, and the available I/O address must be considered in order to assure that the image acquisition card will coexist properly with any other boards on the bus. Finally, image acquisition cards usually have a considerable amount of memory (the frame store) and sometimes this can be mapped into the address space of the host computer. If this is true, care must be taken to see that the base address and size of this memory are compatible with the host computer's memory system.

Another decision that must be made when selecting an image acquisition card is whether the system will capture grey-scale (commonly called black and white) or color images. Each of these formats are discussed further in section 5.6. Grey-scale frame grabbers are generally much less expensive than their color counterparts, so color should only be used if the additional expense is justifiable.

Another specification to consider is resolution. Level resolution is usually specified as the number of bits in the A/D converter and it determines the number of unique intensity levels that can be digitized. A card with 8-bit level resolution can represent 256 different intensity levels, for example. Spatial resolution must also be considered. This specification indicates the number of **pixels** or picture elements into which each image frame will be divided, and it is usually stated as a horizontal and vertical size. Spatial resolutions of 512 × 512 and 640 × 480 are common.

The composition of the incoming video signal must also be considered, and there are two common standards. The RS-170/NTSC standard is used in the United States and is based on a base frequency of 60 Hz. The

CCIR/PAL standard is common in Europe, and it uses a 50 Hz base frequency. The two standards are not compatible, but some frame grabbers are capable of switching between the two for maximum flexibility. The frame grabber and the camera in use must be operating on the same standard for proper image acquisition to take place. The number of video inputs may also be variable. Many cards are available with two or more inputs, and the image-capture hardware can be switched from one input to another via hardware or software control. External video multiplexors can also be used to effectively increase the number of inputs, where one of several video inputs can be selected for transmission to the board.

Image acquisition speed is another specification that is important in inspection applications. The number of frames that can be captured per second is usually the limiting factor on the speed with which inspections can be accomplished. A standard television signal is **interlaced**, which means that odd-numbered lines are transmitted in one frame and even-numbered lines are transmitted in the next frame. An RS-170/NTSC system transmits 60 frames per second, but because it takes two successive interlaced frames to complete an image the actual frame rate is 30 per second. Similarly, European systems typically transmit complete frames at 25 per second. Many commercially available frame grabbers are capable of capturing images at these rates. If higher rates are needed, special purpose cameras and interface boards must be considered. If lower rates are acceptable, the image acquisition time stated by the frame grabber manufacturer must be considered in order to determine a product's suitability.

A word of warning on speed specifications. Although many frame grabbers can achieve 25–30-frame per second acquisition, most busses will not support the transmission of images at these rates. The ISA personal computer bus, for example, has a maximum data transfer capability of approximately 2 megabytes per second (Stalker, 1995), which translates to fewer than 10 image frames per second in practical applications. In other words the computer cannot receive and process the images at the rate that they are acquired, so many frames are lost. This bus bottleneck must be considered when inspection rates are estimated. In order to eliminate this bottleneck, many manufacturers have increased the on-board processing capabilities of their frame grabbers or designed auxiliary image processors that are directly linked to the frame grabber. The capabilities of these processors must be carefully analyzed in order to determine what portion of the image processing required can be done prior to the bus interface. The development of faster busses is also helping to eliminate this bottleneck. The PCI personal computer bus, for example, operating in conjunction with a 120 MHz processor is capable of transferring 80 metabytes per second to the host computer memory (Stalker, 1995), enabling full 30-frame per second acquisition and processing on a properly equipped personal computer.

Synchronization is another issue which often must be considered along with speed. Some form of external triggering or synchronization must be present on the board if image acquisition is to be triggered by external events. This is often the case with inspection systems, where the arrival of a product triggers image capture or where synchronized snapshots are used to acquire images of continuously moving products. Software triggering can also be used, where the host computer triggers image acquisition.

The quality of the captured image is determined by other specifications. Noise can be present in the video signal, or it can be introduced during the digitization and transmission process. Some sort of filtering must be present in order to compensate for this, or noise will propagate into the stored images and aggravate further processing. Linearity is another factor affecting image quality, and it indicates how consistent the digitization process is over the image area. In a simple case, linearity can be determined by checking if a straight line in the scene is indeed straight in the captured image. Aspect ratio is another important image attribute. A true 1:1 aspect ratio in the image acquisition system will cause a circle in the scene to generate a true circle in the image. Anything else will cause the circle to be distorted into an ellipse or worse. If measurements are to be made in the image, linearity and aspect ratio specifications must be checked closely to insure that horizontal, vertical and angular measurements will be comparable.

The processing and storage capabilities available on the image acquisition board are also important. On-board processing can range from simple LUTs to advanced image processing, with an accompanying increase in cost. The size of the frame store is also variable, and often additional frame store memory can be purchased as an option. The frame store size is usually specified as the numbr of full frames that can be stored on the board. If the on-board processing is to involve the comparison or combination of two or more images, the frame store size must be adequate to accommodate this.

Image acquisition boards also differ in their display capabilities. In the simplest case the board has no display capabilities whatsoever, so captured images must be transferred to the host computer and displayed on the monitor there. Other boards include color or grey-scale display capabilities and generate video signals suitable for display on standard monitors. Displays of this type allow the user to view the image that is captured in the frame store and are useful to demonstrate what is happening or to troubleshoot systems. As was stated earlier, many boards incorporate output LUTs in the display circuitry to add emphasis or colors to the output image. Some boards also have a memory area dedicated to overlays where text, graphics, or other information can be overlaid onto the frame store image for display purposes without actually altering the stored image itself.

Software support is another important aspect to consider when purchasing or specifying a frame grabber, and there are three things to look for. First, the manufacturer usually supplies a set of software interface tools and drivers that allow users to write programs that interface with the board. Tools are usually included for such tasks as checking board status, changing modes of image acquisition or display, triggering image acquisition, transferring all or part of the image in the frame store to the host computer, and writing data from the host computer into the frame store. Second, manufacturers often offer complete applications for various computing environments that take care of board interfacing and also perform tasks such as image acquisition, display and saving to disk. Some image processing routines are usually supplied as part of these applications as well. Finally, many image acquisition boards are supported by third party software, so it may be possible to use an application from a separate vendor that fits the requirements of the project particularly well or that a system designer is already familiar with. The level of third party software support for a product is usually a good indication of its popularity in the marketplace.

Listed below are two companies that each offer a wide variety of image acquisition cards. This is not in any way an advertisement for or endorsement of their products, but rather this information is provided as a starting point in the process of gathering information on various alternatives.

Data Translation, Inc.
100 Locke Drive
Marlboro, MA 01752–1192 USA
Phone: 1–800–525–8528

Imaging Technology, Inc.
55 Middlesex Turnpike
Bedford, MA 01730–1421 USA
Phone: 1–800–333–3035

Each of these companies also has sales offices in Europe and elsewhere.

And now a final warning on image acquisition cards. The recent explosion in interest in multimedia personal computing has caused many companies to develop cards that will interface to video signals. These cards are typically low in cost (less than $500) and, although they are primarily designed for reading motion video into a computer, many of them have a capability to store individual frames. At first glance these boards may seem to be an ideal low-cost alternative to sophisticated frame grabbers, but buyers should beware of their shortcomings. Because the emphasis is on motion video and low cost, these cards often do a poor job of capturing individual frames. Single images are often blurred or very noisy, and if proper synchronization is not used the image may in fact consist of parts of two consecutive frames and may appear torn, due to motion that occurred between the frames. Linearity and aspect ratio are also suspect with some of these boards, so accurate measurements and sophisticated techniques such as object recognition are likely to be difficult or impossible. On-board processing is also very limited, and it usually emphasizes video compression algorithms rather than image processing. In short, with frame grabbers, as with most other things in life, you get what you pay for. It is worthwhile to keep watch on this market, however. As these multimedia boards become more popular, manufacturers can take advantage of economies of scale and make their products more sophisticated while still keeping the price down. Home video cameras provide a good analogy: the cameras now available for less than $1000 are more sophisticated than the multimillion dollar imaging system that NASA landed on the planet Mars.

5.5 IMAGE PROCESSING HARDWARE

5.5.1 Personal computers

The most common personal computers in use today are based on the Intel architecture, and are still called IBM compatibles even though that is now somewhat a misnomer. The two most common processors being sold at the time of this writing are the Intel Pentium and the Intel 486. Pentium-based personal computers represent a level of processing power and speed previously unavailable in the personal computer market, and they rival the processing capability of more expensive workstations.

Although virtually all of these PCs utilize the ISA expansion bus architecture, many of them also implement a high speed local bus to accelerate I/O-intensive tasks such as disk access and graphics output. Two common local bus architectures are the VESA standard and the PCI standard. Frame grabber boards are available for each of these local bus architectures. Very fast image acquisition, transfer and processing can be achieved by taking advantage of the combination of a fast local bus like PCI and a powerful processor like the Intel Pentium.

The most popular operating systems for personal computers are

Microsoft's windows-based system Windows 3.1 and Windows 95. These operating systems put an emphasis on multimedia applications that demand high speed input and output of video and graphical information. Because these operating systems are graphics-based (rather than text-based like MS-DOS), the functions required for image processing and display are much easier to implement. This has resulted in a proliferation of image processing programs. A wide variety of such programs are available for a reasonable cost, and some good programs are even available as shareware for less than $50.

Two other personal computer architectures deserve mention as potential image processing computers, the Macintosh and the Power-PC. Macintosh computers have for a long time been preferred in the areas of image processing and graphical display. Programs written for the Mac seem to be easier to use and more powerful than their Windows-based counterparts, although this playing-field may be more level with the recent release of Windows 95. The problem with Macintosh computers is that their small market share does not justify the kind of development expense necessary to generate complex applications like image processing programs, so fewer applications are available to choose from.

The Power-PC architecture represents a true hybrid of personal computers and workstations. Although it is essentially a personal computer architecture, the Power-PC employs a processor, similar to those used by many workstations, that uses RISC (reduced instruction set computing) technology. This makes the Power-PC architecture inherently better for many image processing tasks, and it may find a niche market in such applications. The Power-PC presently has limited availability and a tiny market share, so software and hardware development are in their infancy and this limits its appeal. The development of the Power-PC is worth watching, however, because of its excellent potential in the area of image processing.

In short, the recent developments in the personal computer industry have played right into the hands of inspection system developers. Many excellent hardware and software alternatives are available for a reasonable cost. It is recommended that, when selecting a system for purchase as a development system or as part of an operational inspection system, the best available combination of processing speed and local bus architecture be chosen. Although this may represent a higher cost than is deemed necessary, such a system will have a much longer useful life than a minimal system that will soon be obsolete.

Emphasis should also be placed on disk storage and graphical display capabilities when specifying a computer for use in image processing. Images require large amounts of storage, so fixed disk sizes should be as large as possible. Also, high quality monitors result in much better graphical displays and higher resolution, two attributes that are valuable when evaluating or processing images.

5.5.2 Workstations

The traditional workhorses of the image processing domain are UNIX-based workstations. Because they have long had the processing power and speed only recently achieved by PCs, they were the obvious choice for I/O- and processor-intensive tasks. Most workstations utilize RISC processors which take advantage of a type of parallelism called **pipelining** to perform more than one task at a time. Pipelining and parallel processing are explained in the next section. Workstations commonly use some form of the UNIX operating system. There are many manufacturers of workstations, but two popular vendors are Sun Microsystems and Hewlett-Packard. Most serious image processing research is still done on workstations.

Workstations have many drawbacks, however. The UNIX operating system has not been developed as rapidly as personal computer operating systems, so it is still very cryptic and unfriendly to the user. Expense is another problem, because the relatively small workstation market has not allowed manufacturers to take advantage of economies of scale to the extent that they have in the personal computer market. Both hardware and software are often several times the cost in a workstation environment when compared to PCs. Workstations are only recommended when extremely high performance is a necessity and cost and usability are secondary issues.

5.5.3 Parallel computers

Most computers in use today still use a basically serial design. In this design a task is not started until its predecessor is completed, and most programming languages are still based on this assumption. Lines of code are written serially, and they are expected to be executed in order. Many image processing tasks, however, can be performed in parallel. Whenever the same operation must be performed on many small areas of an image, and these operations are not dependent on each other (i.e. the result of one operation is not used as input to another), these operations can be executed simultaneously. Many low and intermediate level image processing tasks can be done in parallel, and using artificial neural networks for processing or recognition increases this advantage since they are also inherently parallel in their design.

As previously mentioned, pipelining is one approach to doing more than one task at a time. Pipelining involves dividing each instruction into subtasks of approximately equal size, and executing subtasks of multiple instructions simultaneously. For this to happen each subtask must use a different part of the processor, and the RISC design philosophy guarantees that this occurs often.

Another approach is true parallelism, where there are multiple processors each executing individual instructions simultaneously. This approach puts fewer constraints on the instructions, since processor resources are not shared. In a true parallel system, if there are 16 processors, 16 instructions can be executed at once. Problems are encountered in parallel systems, however, whenever operations depend on each other or when a great deal of communication is necessary between processors. Consider, for example, a situation where 16 numbers are computed by simple addition and then the maximum of the numbers is to be found. The 16 addition operations are easily conducted in parallel, but then the results must be communicated so they can be compared. Such communication often requires more time than the initial computation.

Both pipelining and parallelism can be demonstrated by using a fast-food restaurant as an example. A serial computer would be represented by a single cashier who takes the order, receives the money, and delivers the food. One customer at a time is served, and no other customer is served until the current customer has a meal.

A drive-through line represents the pipelining approach to improving throughput. The task of serving a customer is divided into three approximately equal parts, each different: order taking, money receipt and food delivery. A customer places an order at one window, pays for the meal at the next window, and receives the food at a final window. The improvement comes because while one customer is paying, an order is being taken from the next customer in line. Although the time required to serve a single customer has not improved, the throughput (i.e. the number of customers served per hour) has tripled. The term pipelining is used because computer instructions, like the customers in the drive-through, are in a pipeline, one behind the other and progress through the processor at the same rate. In an RISC processor, instructions are usually divided into subtasks, such as fetch instruction from cache, fetch operands from registers, perform operation and write result to register.

True parallelism is represented by the cash registers inside the fast-food restaurant. One customer is served at a time by each cashier, but there are several registers operating in parallel. The improvement in throughput is proportional to the number of registers operating, so if three cashiers are working the number of customers served per hour is tripled. The advantage of true parallelism is that it places fewer restrictions on the instructions. Instructions need not all consist of the same approximately equal subtasks. If it takes twice as long to take an order from a specific customer inside the restaurant, for example, the throughput at the other registers is not affected. If this happens in the drive-through line, however, the entire pipeline is held up and throughput is reduced at all stations.

Parallel computers for image processing are becoming more common and less expensive. Pipelining is used in most RISC workstations and high-end PCs and it is well suited to image processing tasks. Multiprocessor computers that utilize true parallelism are also becoming more common. Because image processing is well suited to parallel computation, several parallel computers have been designed specifically for this purpose. Some companies even manufacture parallel processing expansion boards for personal computers. If an inspection system required the speed of parallel computation, for example, the system could be developed on a workstation linked to a parallel computer, and then loaded onto an expansion board in a PC to make an affordable final package (Adaptive Solutions, 1995).

A major drawback of parallel computers is the current lack of standardized parallel programming languages. Each manufacturer often develops a language that is product specific, so much time is spent becoming familiar with the language and its application. Also, people who have performed serial programming all of their lives have a difficult time 'thinking in parallel', as is required in multiprocessor systems. Programming in parallel requires an entirely different approach than serial programming, and many familiar serial techniques and data structures are invalid in the parallel environment. In short, expect programming and development time to be long if an unfamiliar parallel computer is to be used. As in the case of workstations, parallel computers should only be employed when the speed or complexity of processing required justifies the additional difficulty and expense.

5.5.4 Special-purpose hardware

Many image processing tasks are well defined and involve the repeated use of techniques like filtering and convolution. Because of this, specific equipment has been designed to accomplish many image processing tasks in hardware. A hardware-specific design allows tasks to be accomplished at maximum speeds because the design is optimized for the task it is intended to perform. Integrated circuits have been designed for tasks like image compression, noise filtering, edge detection and even pattern recognition.

In addition to high processing speed, special-purpose hardware often represents a cost advantage. Once a chip is designed, including it in an image processing system is relatively inexpensive. The disadvantages of special-purpose hardware are inflexibility and development time. Although many adjustable parameters are often included in the design of a chip, it always seems that the processing required for a specific application does not quite fit any available standard product. The alternative is to develop a chip design specifically for the task of interest, and this is obviously not justified unless the number of potential applications is very large. At present, special-purpose image processing hardware is limited to high volume or research and development applications.

5.6 IMAGE FORMATS

5.6.1 Introduction

The standard format for a digitized image is a rectangular array of discrete points called picture elements or pixels, with the value of each pixel representing the light intensity at that point in the original scene (Hollingum, 1984). The spatial resolution of an image is determined by the number of pixels that are used to represent the image. The larger the number of pixels, the smaller the grain of the representation and the higher the resolution and accuracy. Some common array sizes are 256 × 256, 512 × 512 and 640 × 480. High resolution comes at a price, however. A single 640 × 480 image, stored in an 8-bit format (one byte per pixel), requires 300 kilobytes of storage space. Figure 5.4 illustrates the effects of spatial resolution. An image of former American President Abraham Lincoln is shown at spatial resolutions of 120 × 160 pixels, 30 × 40 pixels, and 12 × 16 pixels.

5.6.2 Binary images

The simplest image representation requires only one bit per pixel, either on or off. Images of this type are often called **binary** images, since pixels can take on only one of two values, for example 1 for white and 0 for black. Binary images are usually generated by passing the original signal through a simple threshold detector, with signals above the threshold generating a 1 and those below generating a 0. The threshold value is often adjustable. Figure 5.5 shows a binary image of a pair of scissors.

Binary images are particularly useful when backlighting is used and object silhouettes are extracted. Backlighting generally makes it easy to extract the object silhouette accurately with a simple threshold function. The resulting binary image can be used to make measurements by counting the number of black pixels in a line, or to compute the area of the object by counting the total number of black pixels. Similarly, the size, shape and location of any holes in the object can be determined.

Figure 5.4 Images of Abraham Lincoln shown at three different spatial resolutions (120 × 160, 30 × 40 and 12 × 16).

Figure 5.5 Binary image of a pair of scissors.

Geometric properties of the object, like the location of the centroid of the object or the orientation of the object's major axis, can also be extracted easily from a binary image.

5.6.3 Grey-scale images

One system where pixels are each represented by more than one bit is called **grey-scale** representation. In this sytem each pixel is represented by a binary byte or word, the value of which represents the light intensity at that point in the image. The number of bits per pixel determines the number of different intensity levels that can be represented. Two common representations are 4 bits per pixel (0–15) and 8 bits per pixel (0–255). Grey-scale images obviously provide much more information about a scene than binary images do, and a high resolution grey-scale image resembles a black-and-white television picture. Figure 5.6 shows a grey-scale image of the same pair of scissors shown in the previous figure.

Grey-scale images provide much more information than binary images, since varying shades in the original scene can be represented. Note that from the additional information provided by this format it could be determined from the grey-scale image of the scissors if the handles were painted properly or if the center fastener was present, each of which would be impossible to determine from the binary image.

Grey-scale images are currently used most often in inspection systems. This is probably because they represent a good compromise between system cost and performance when compared to binary or color systems. A wide range of processing techniques are also available for grey-scale images. Grey-scale images have been the subject of much research over many years, the result of which has been the development of many useful, well-developed processing techniques. Due to reductions in cost, the balance is starting to shift toward full-color systems, however, and these will soon be as popular as grey-scale systems.

Figure 5.6 Grey-scale image of a pair of scissors.

5.6.4 Color images

Color image representations preserve even more of the information present in a scene. Color images usually consist of three separate image representations called **color planes**, with pixel values in each plane corresponding to the intensity of a certain color at a specific point in the image. Each color plane consists of an array of pixel values similar to the grey-scale representation. The most popular system is called RGB (red–green–blue), where three color planes are used to represent the intensity of red, green and blue in the scene (Levine, 1985). By combining the values of the three planes, any true color can be represented. A common representation is 24-bit color, where the red, green and blue values for a pixel are each represented by 8 bits. Over 16 million different colors can be represented in this format.

There are obvious uses for color in inspection systems. Many products are color coded, and could be easily recognized with a color system. Inspecting printed and painted products also often requires color images. The inspection of agricultural products for features such as ripeness and bruises is usually dependent on color, and many medical applications involve judging skin or tissue color. The drawbacks of color are the initial expense of color cameras and frame grabbers, and the storage required for color images. An RGB image with 8-bit level resolution requires three times as much storage space as a grey-scale image of the same scene.

5.6.5 Format conversion

Conversion between different image representation formats is often desired. Color (RGB) images are typically converted to grey-scale by using the intensity formula in equation 5.1 (Levine, 1985):

$$I = 0.612 \times R + 0.369 \times G + 0.019 \times B \tag{5.1}$$

This calculation is done for each pixel, with the result being rounded off to the desired number of grey-scale bits. Grey-scale images can be converted to binary format by thresholding, where each pixel is replaced by a 1 if it is above the threshold or by a 0 if it is below. Grey-scale images are also occasionally converted to color images, by choosing different colors to represent the different intensity values. These images are referred to as **false-color** images, since the colors do not represent those present in the original scene.

5.6.6 Image transmission and storage

An unprocessed image received directly from a frame grabber is often called a **raw** image. A raw image is usually stored as simply a rectangular array of pixel values. A 640 × 480 grey-scale image stored in 8-bit resolution will require 640 × 480 × 1 = 307 200 bytes of storage. A one-second sequence of such frames acquired at video rates (30 per second) would require over 9 megabytes of storage. If the images are 24-bit RGB color, the storage requirements are tripled.

Transmission of images across computer networks poses similar challenges due to the size of image files and the limited capacity of networks. A typical local area network, for example, would overload attempting to transmit a single video-rate image stream. Wide area networks are typically much slower. Transmitting a single digitized image over the internet, for example, can take several seconds.

Fortunately, image files can often be compressed to a fraction of their original size. A simple compression technique involves making the image actually smaller. A 4:1 compression can be achieved, for example, by simply replacing each 2 × 2 area in an image with a single pixel that represents the average of the four replaced. This type of technique is called **lossy compression** because some of the information in the original image is lost. The compressed image cannot be decompressed to reproduce the exact original image.

Other compression techniques are **lossless**, where the original image can be reproduced from the compressed version. Although this may at first seem like getting something for nothing, these techniques take advantage of redundancy that exists in most real images. One such method is called **run length encoding** (RLE). Instead of simply representing an image as an array of pixels, RLE breaks down each line of the image into a series of runs of common color. The color is first stored, then the number of continuous pixels of that color. Images with large areas of the same color can be greatly compressed using this technique.

Differential encoding is also used as a compression technique. Because pixel values vary rather slowly across real images, the number of bits required to represent the difference between adjacent pixels is often much less than the number of bits required to represent the absolute pixel value. This can be taken advantage of by storing only the value of the first pixel in a line of the image, and then storing only the difference between adjacent pixels.

Color reduction is another compression technique. The number of bits used to represent the absolute pixel value in an image will often represent many more colors than are actually present in the image. For example, the 24-bit RGB format that is commonly used can represent over 16 million colors, but most images have much fewer than 16 million pixels! Only a fraction of the available colors would be used even if every pixel was a unique color. Color reduction methods count the colors actually used in an image, create a table of those colors, and use a representation for each pixel that is an index into that table. The index can be stored with fewer bits than the original color, so a smaller file results.

Another approach to image compression is to treat the image like any other file and simply use one of the available data compression techniques. The Lempel-Ziv and Welch (LZW) algorithm is a sophisticated data compression technique that enables one-pass encoding and decoding of any data stream (Welch, 1984), and it is used in both image and non-image applications.

Most image compression techniques in use today are much more sophisticated than those just described, but the principle of eliminating redundancy is the same. Many techniques are in fact a combination of methods, where different compression approaches are used one after the other. The GIF (graphics interchange format) technique, for example, uses color reduction followed by LZW compression.

I have conducted a simple experiment in order to illustrate the effectiveness of certain compression techniques. First, three images were saved in a raw rectangular (320 × 480) format. These images were then compressed using the GIF compression technique and a technique developed by the Joint Photographic Experts Group (JPEG), and the resulting file sizes were compared to the original raw file sizes. The images used were a real image of a soda can, a random image consisting of totally random pixel values, and a plain image where all of the pixels were of the same value. The results are shown in Table 5.1.

Table 5.1 Bytes of storage required for different image formats

Image format	*Soda-can image*	*Random image*	*Plain image*
RAW	153,600	153,600	153,600
GIF	94,334	126,520	1,461
JPEG	57,298	201,156	2,131

The GIF format is a lossless technique that can reproduce the original image exactly. The JPEG technique is a lossy method based on a transformation called the discrete cosine transform, and the degree of loss can be adjusted to suit specific uses. A low loss version of JPEG was used here to generate comparable results.

Note that the two artificial images represented the ends of the spectrum. The random image was compressed only slightly by the GIF technique and was actually expanded by the JPEG technique. This is because the compression technique generates certain overhead that must be stored in the file, and the compression did not reduce the image adequately to even justify the overhead in the JPEG case. The plain image, on the other hand, was reduced in size by approximately 99% by each of the compression techniques. Obviously, there is a large amount of redundancy that can be removed from this image, since all pixels have the same value. The results with the real (soda can) image are probably the most meaningful, however. Here the GIF technique reduced the file size by 39% and the JPEG technique reduced the file size by 63%.

6

Low-level image processing

6.1 INTRODUCTION

An image that is originally acquired by a camera is called a raw image, and raw images must be processed in some way to extract the useful information that they contain. The first step in this procedure is called **pre-processing**, and it is done to prepare the image for further processing at higher levels. The two primary goals of image pre-processing are to eliminate unwanted attributes in the image that would aggravate subsequent processing steps, and to extract primitive features that represent the pertinent information present in the image. Unwanted or undesirable image attributes include noise, areas of featureless space, or areas where the features can be ignored. Primitive features include edges, corners or vertices, lines and line junctions. These features typically represent the boundaries or surface details of objects that are valuable when those objects are to be located and identified in an image.

Several useful pre-processing techniques are discussed in the sections that follow. Each of the techniques addresses the goals of pre-processing in one way or another, and they usually have another benefit as well. Pre-processing has an information-filtering effect. Of all the data that is represented by the array of pixel values, only a small percentage is actually meaningful. Pre-processing sorts the useful data from the unnecessary clutter, extracting only the information that represents important features of the objects present in the image. After pre-processing, an image can often be represented by much less data, and each piece of data retained has much more meaning than the simple raw pixel values.

6.2 IMAGE SIZE REDUCTION

It is often desirable to reduce the number of pixels in an image. A reduction in pixel count, or image size, makes subsequent processing faster and reduces storage and transmission requirements. Images are often of adequate scale to allow size reduction without making processing or recognition more difficult, and many times the reduction process will actually eliminate unnecessary detail or unwanted noise. Three common approaches to image size reduction discussed next are cropping, sampling and averaging.

Cropping is the simplest form of image size reduction. Cropping simply eliminates unwanted or unnecessary portions of the image by cutting them away. A typical approach to cropping is to roughly locate the area of interest in the scene, and then to construct a rectangle of appropriate size around that area. The pixels inside the rectangle are retained and those outside are discarded. In many practical applications the location of an object of interest in the image is at least approximately known, and often the object does not occupy the entire image area. A simple cropping procedure can significantly simplify processing, storage, and transmission problems.

The second size reduction technique is applied over the entire image area and it is called **pixel averaging**. With this approach an image is usually divided into many regions of equal size, and each of the regions is replaced by a single pixel that represents the average intensity value of the region. For example, each 2×2 area in the image could be replaced by a single pixel representing the average of the four original pixels, resulting in a 4:1 compression of image size. Averaging usually reduces the range of brightness values in an image, so it has a tendency to eliminate some details. This also has the beneficial effect, however, of eliminating unwanted noise.

The final image size reduction method is also applied over the entire image area and it is called **sampling** or **resampling**. Because an image is formed of discrete pixels, it can be thought of as a collection of samples of the brightness in a scene. One way to reduce the number of pixels is simply to sample the image at a reduced density. Sampling the upper-left pixel of each 2×2 region of the image, for example, would result in a 4:1 reduction in image size. This is often called resampling because in effect a sampling is being taken of the image which is itself a sampling of the original continuous scene. Sampling treats all areas of the image equally and, unlike averaging, it preserves the full range of brightness values in the original scene. In this way it preserves the maximum amount of detail, but it also allows noise to be transferred into the reduced image.

Figure 6.1 illustrates the effects of sampling and averaging as size reduction techniques. The figure shows an image of a deer with 10% added noise as the original image. The two reduced images were produced by sampling and averaging over a 2×2 area as described above. Note that the sampled image preserves more detail but also

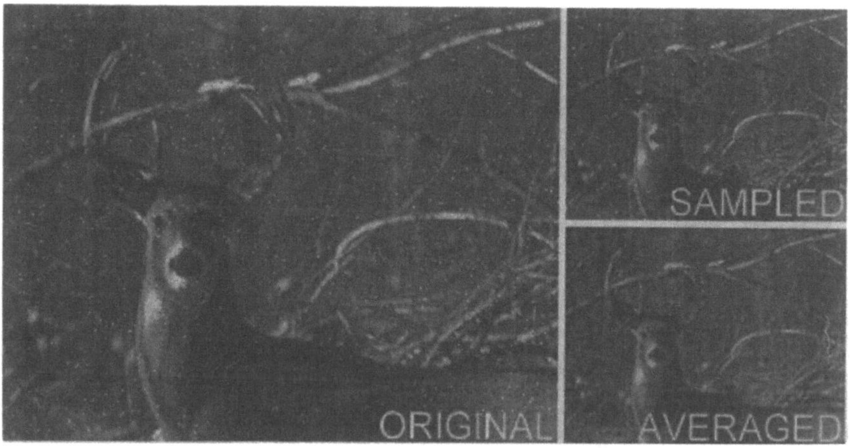

Figure 6.1 The effects of the size reduction techniques of sampling and averaging on a noisy image of a deer.

contains approximately the same level of noise as the original, whereas the averaged image shows slightly less detail but has a greatly reduced noise level.

6.3 NOISE REMOVAL AND FILTERING

Some degree of noise is present in most real images, and it is often desirable to remove it. A small number of randomly located pixels are often much brighter or darker than their neighbors, for example, and this is referred to as **salt-and-pepper noise**. This type of noise can be eliminated by simple filtering, where a neighborhood of pixel values around a given pixel are allowed to influence its value. An example of this is simple averaging, where a pixel takes on the average value of its neighborhood. Although averaging filters remove undesirable properties like noise, they also tend to blur the image, making some desirable features less distinct.

Several other filtering techniques are also often used in image preprocessing. Smoothing techniques are similar to averaging, but a weighted average is computed instead. The value of the pixel being replaced is given more weight than neighboring pixels, so less detail is lost while noise is still removed. Detail-enhancement filters have the opposite effect. The value of the pixel being replaced is given significantly more weight than neighboring pixels, and neighboring pixels may even be given negative weight to de-emphasize them. Figure 6.2 shows the results of some filtering techniques on the deer image

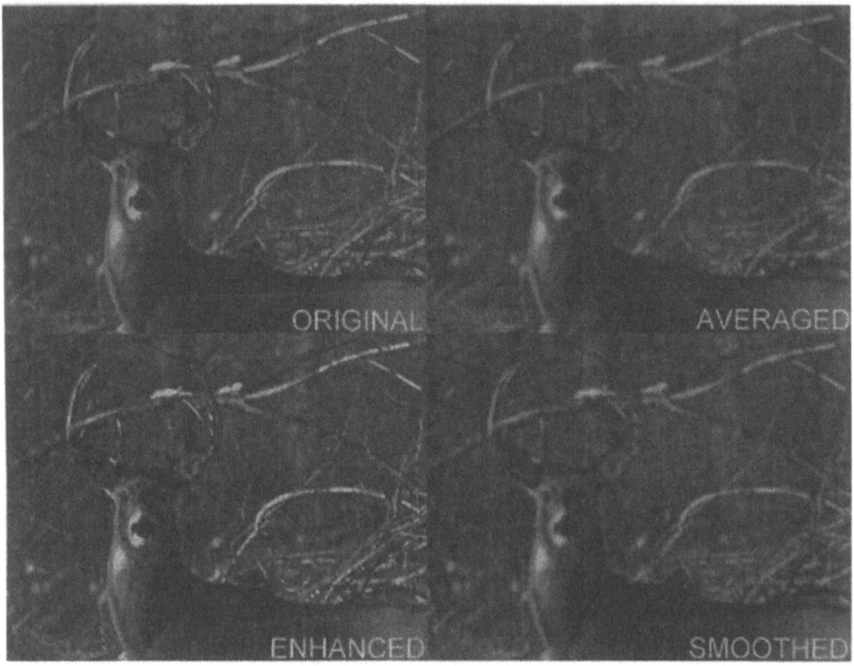

Figure 6.2 The effects of filtering techniques for averaging, smoothing and detail-enhancing on an image of a deer.

shown previously, this time without added noise. The figure shows the original image, and images created by averaging, smoothing and detail-enhancing the original.

6.4 THRESHOLDING AND HISTOGRAMS

6.4.1 Thresholding

Another popular pre-processing technique is thresholding. As described earlier, thresholding allows the number of discrete intensity levels in an image to be reduced. Application of a single threshold results in a binary image, but multivalued thresholds can also be used to reduce intensity resolution. For example a grey-scale image of 0–255 resolution can be thresholded at various levels to reduce the intensity resolution to 0–15 for display on a personal computer monitor that does not support the higher resolution.

Thresholding is usually used to differentiate objects of different colors in an image, or to separate objects from the background, and the primary problem is the determination of the proper threshold values to

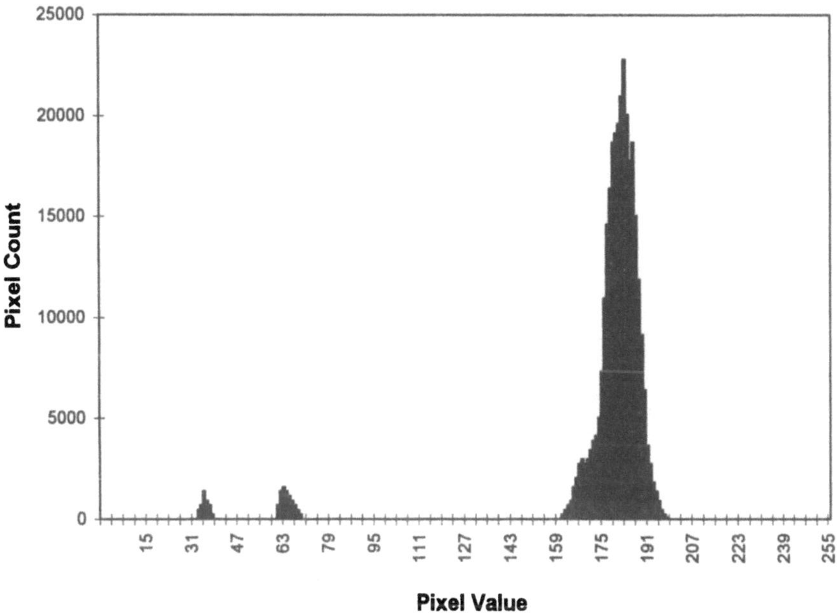

Figure 6.3 Histogram showing three distinct brightness peaks.

use. To separate an object from the background, for example, the threshold should be set between the average intensity of the object and the average intensity of the background. In order to determine this point, a **histogram** of the image is often required. A histogram is a graph of the frequency of occurrence of each level of intensity in the image. This graph should consist of easily separated peaks if the intensity difference between the object and the background is adequate (i.e. the image is a high-contrast image). This fact can be clearly seen in Fig. 6.3, which is a histogram for the grey-scale image of a pair of scissors that was shown in Fig. 5.6. The proper threshold value falls in the trough between the two dark pixel peaks and the peak representing the lighter background pixels (in this case approximately 112). The binary image shown in Fig. 5.5 is actually the result of applying this threshold.

6.4.2 Histogram equalization

Another image processing technique based on histograms is **histogram equalization**. Image intensities are usually unevenly distributed in a histogram, as was the case in the previous example (Fig. 6.3). Histogram equalization seeks to level the histogram so that approximately the same number of pixels of each intensity are present in the image. This is accomplished by mathematically compressing intensity regions where

Figure 6.4 A relatively dark image containing handwritten letters.

Figure 6.5 The result of applying histogram equalization to the image of Fig. 6.4, making the letters more distinct.

few pixels are represented into fewer intensity levels, and expanding those where many pixels are represented into more levels. Histogram equalization has the effect of increasing the contrast in images that are dominated by dark or light pixels. In the image of Fig. 6.4, for example, it may be desirable to extract the handwritten letters. Although they are faintly visible, the preponderance of dark pixels in the image makes them indistinct. Applying histogram equalization to this image makes the letters much easier to extract, as can be seen in Fig. 6.5.

6.5 REGION GROWING AND HOLE FILLING

6.5.1 Connectedness

Although the presence of regions of differing intensity in images is often obvious to humans, it is not to a computer. To a computer vision system an image is simply a rectangular collection of individual pixels. The fact that a group of pixels may be similar in intensity and also spatially proximate is something that must be explicitly determined. The first step in accomplishing this is to define what constitutes a connection between two pixels.

The two types of connectedness that are usually defined in a rectangular grid like an image are called **four-connectedness** and **eight-connectedness** (Horn, 1986). If the analogy of a compass is used, a pixel is four-connected to its neighbors to the north, south, east and west. Using the same analogy, a pixel is eight-connected to the original four and also to its diagonal neighbors to the northeast, northwest, southeast and southwest. The type of connectedness that is applied determines how regions will grow in an image. Four-connectedness and eight-connectedness are illustrated in Fig. 6.6.

6.5.2 Region growing

A region is usually defined as a group of connected pixels that are also similar in intensity. The simplest form of region growing begins at a seed pixel and proceeds by examining all of the pixels immediately connected (either four-connected or eight-connected) to the seed pixel. Those that are similar enough to the seed pixel are included in the region. The region-building process continues by investigating the pixels connected to all of the new pixels added to the region, and this process continues iteratively until no similar connected pixels are found. The most common example of this type of region growing is the 'flood-fill' command usually found in simple computer drawing or painting programs.

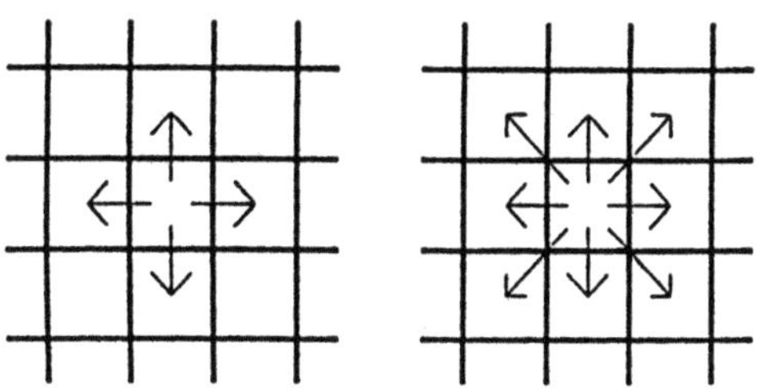

Figure 6.6 Four-connectedness (left) and eight-connectedness (right).

It is often desirable to identify all of the regions present in an image by assigning each of them a unique region number. The simplest way to do this is to simply grow the regions serially, one at a time. The region number can be set to an initial value, any initial seed point can be used, and the region-growing process can be applied as just described. Each new pixel added to the region is assigned the current region number. When the first region is complete, the image can be searched for any pixel not yet assigned a region number and that pixel can be used as the next seed point. The region number is incremented, and the region-growing process begins again at this new seed point. This entire procedure is continued until no pixel is without a region number assignment.

Because the serial region identification is time consuming, faster methods have been developed. One of these is called simultaneous region growing. With this method several seed points are chosen and many regions are grown simultaneously. An extra step is introduced into the region-growing process to check if any similar pixels encountered have previously been assigned a region number. If so, the equivalence of these region numbers is noted so that two or more simultaneously grown connected regions can be joined into a single larger region once the procedure has terminated. Seed points can be located randomly or at evenly spaced points in the image. Simultaneous region growing is especially attractive in a parallel computing environment where its application can result in a significant reduction in processing time.

6.5.3 Hole filling

A hole is defined as a region contained within another region. In many applications, holes represent undesirable surface features such as noise, shadows or glare, and it is desirable to fill them prior to further processing. A technique similar to region growing is usually used to fill holes. Instead of assigning region numbers as described previously, the pixels in the hole are assigned intensity values consistent with the surrounding region. Most hole-filling algorithms require the hole to be below some maximum size before it will be filled. This allows the removal of unwanted features without filling in areas that may represent true holes in the object. Consider, for example, the image in Fig. 6.7. This image represents an object with four symmetrically spaced holes, but also with some noise or speckle present. A hole-filling algorithm that fills all holes of three pixels or less in size will neatly remove the noise without affecting the real holes in any way, as can be seen in Fig. 6.8.

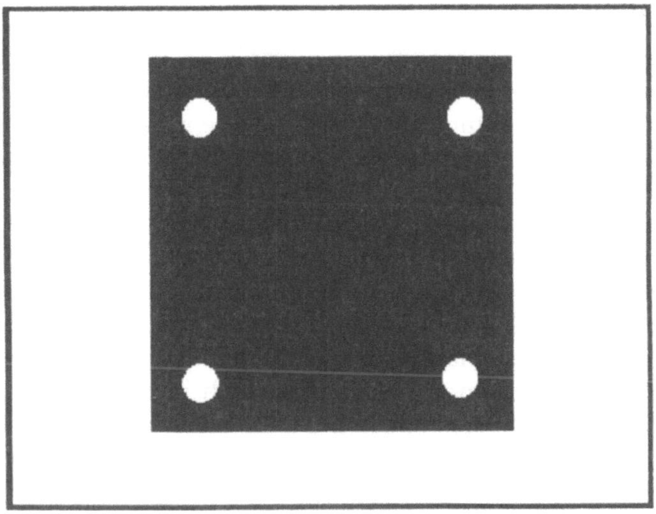

Figure 6.7 Image of an object with four holes which are meaningful features and some noise speckles which are not.

Figure 6.8 The result of applying size-selective hole filling to the image of Fig. 6.7.

6.6 EDGE DETECTION

Edge detection is one of the most popular image processing techniques. Edges are defined as lines along spatial discontinuities in intensity in the image. These discontinuities usually represent the outline of objects, surface features on objects, or lines in the image. Once an image is edge detected, the ideal result is a line-drawing representing all pertinent edges in the original image. This is often difficult to achieve in practice, however. The presence of noise, glare, shadows and other distractions often results in missing edges or in extraneous edges appearing.

There are two main approaches to edge detection, the computational approach and the template technique. In the computational approach, a value of some function is calculated at each point in the image. The function is usually dependent on the values of the pixels in the immediate neighborhood surrounding the point. The most popular function for the computational approach is called the Laplacian-of-Gaussian (LoG) (Marr and Hildreth, 1980). The Laplacian operator, which takes the second derivative in two dimensions (equation 6.1) is applied to the two-dimensional Gaussian function (equation 6.2).

$$\nabla^2 = \frac{\partial^2}{\partial x^2} + \frac{\partial^2}{\partial y^2} \tag{6.1}$$

$$G_{x,y} = \frac{1}{\sqrt{2\pi}\cdot\sigma}\, e^{-\frac{1}{2}\left(\frac{x^2+y^2}{\sigma^2}\right)} \tag{6.2}$$

where x and y are the horizontal and vertical displacements from the center of the Gaussian function.

The resulting operator (the LoG) is shown in equation 6.3:

$$\nabla^2 G_{x,y} = \frac{1}{\sqrt{2\pi}\sigma^3}\left(\frac{x^2+y^2}{\sigma^2} - 2\right) e^{-\frac{1}{2}\left(\frac{x^2+y^2}{\sigma^2}\right)} \tag{6.3}$$

The first and second derivatives of the Gaussian function in one dimension are shown in Fig. 6.9. Note how similar the second derivative response is to the DoG model of ganglion cell response shown in Chapter 3 (Fig. 3.7). Since the LoG operator operates in two dimensions across the image, the shape of the response of the operator is the three-dimensional surface shown in Fig. 6.10. Once again, the response of the LoG is very similar to the DoG biological model presented in Chapter 3 (Fig. 3.8). Once the LoG edge-detection operator is applied to an image,

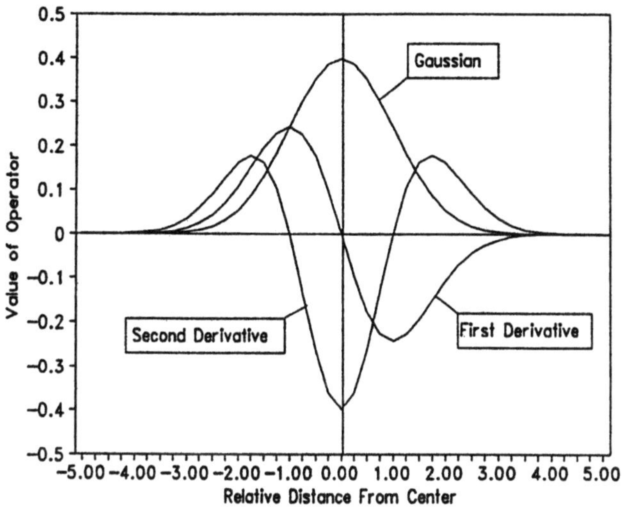

Figure 6.9 The first and second derivatives of the one-dimensional Gaussian function.

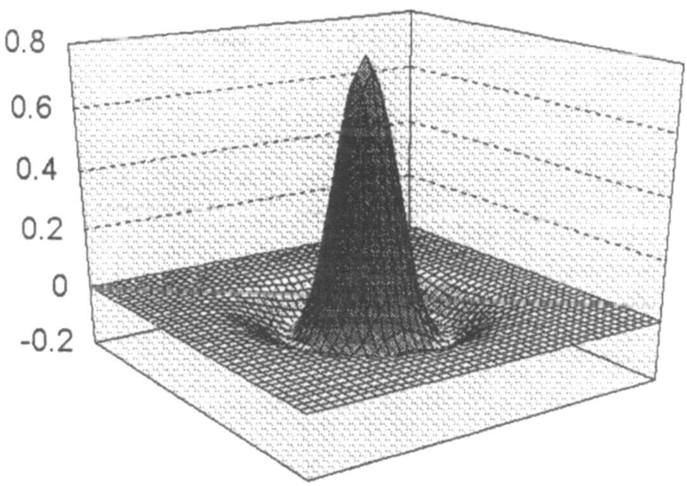

Figure 6.10 The two-dimensional Laplacian-of-Gaussian operator.

each pixel is replaced by its corresponding edge value. The LoG operator gives no indication of the orientation of the edge, only its strength.

The other approach to edge detection is the template approach. With this approach each pixel and its surrounding neighborhood are combined with a rectangular template of values, and the result of that combination represents the likelihood of an edge being present at that point. Two popular edge-detection operators based on templates are the

-1	0
0	1

0	1
-1	0

Roberts Cross

-1	0	1
-2	0	2
-1	0	1

-1	-2	-1
0	0	0
1	2	1

Sobel Operator

Figure 6.11 The templates used on the Sobel and Roberts edge detection operators.

Sobel operator (Duda and Hart, 1973), and the Roberts cross (Roberts, 1965). The templates used in each operator can be seen in Fig. 6.11. Each operator consists of two templates, one for measuring the strength of an edge in each of two orthogonal directions. The edge value generated by each template is computed as the sum of the product of each of the weights shown in the figure with the value of the corresponding pixel, so the operation can be thought of as sliding the chosen template over the image and calculating the degree of match at each point. The values generated by each of the templates are then combined to indicate the overall edge value at the point of interest, usually by taking the square root of the sum of their squares or by simply adding their absolute values. The values generated by each of the templates can also be considered individually as an indication of the orientation of the edge.

The author has tested all three of these edge-detection operators. Neither the LoG nor the Roberts cross operator performed as consistently well on a variety of images as the Sobel operator. The primary difficulty encountered with both the LoG and the Roberts cross was the fact that they tended to enhance undesirable noise in the image, producing many spurious edge responses that did not coincide with perceptually obvious edges. The Sobel template is also computationally very simple and efficient to implement, making it an excellent choice for general purpose edge detection. Results of applying the Sobel edge detection technique to real images can be seen in Chapter 9.

6.7 PRIMITIVE IMAGE FEATURES

It is often desirable to identify and locate certain features in an image. In a manufacturing situation, features such as holes or slots may be used to differentiate parts, or to locate and orient components. Many inspection applications also rely on feature detection, either by identifying desirable features or by recognizing defects. There are two different approaches to feature extraction, one that works with the original image and another that works with edge representations.

One simple method of feature detection that uses original images is template matching. With this technique, areas of an image are compared to a stored template, and the degree of match is calculated. If the match is good enough, the feature is identified. This method works well in controlled situations, where the location and size of a feature in an image is fixed or known, but it works poorly in unstructured cases.

For the more difficult feature identification situations, advanced techniques have been developed that work with edge-detected images. These methods can be insensitive to the location, orientation or scale of the feature, depending on the requirements of the application. Vertices, line segments and junctions have all been used as primitive features in edge images. Vertices are typically defined as corners or points of high curvature along edges, and they have proved to be valuable features for object identification (Attneave, 1954; Biederman, 1985). If the location of vertices along an edge can be determined, and the connections between vertices established, a fairly accurate 'connect-the-dot' type of representation of an edge image can be constructed from a small amount of information. The relationships between vertex locations can also be used to identify objects. Details of a vertex extraction and connection technique are given in Chapter 9.

Line segments are another popular type of image feature that can be extracted from an edge image. Line segments can be extracted by analyzing the edge pixels to determine if groups of them are collinear. Large groups of edge pixels that are nearly collinear can then be replaced by a simple line segment between two points. There are two approaches to line segment extraction, the local approach and the global approach. The local approach examines short edge segments that are near each other in the image and determines if they are collinear. If so, they are replaced by a single line segment. This process is repeated iteratively until typically many small edge segments have been grouped into a few longer line segments.

A popular global approach to line segment extraction uses the **Hough transform** (Marshall and Martin, 1992). This transform determines, for each point in the edge image, the set of lines that could pass through that point. When the transform is applied to all edge points in the image, those that are collinear will generate groupings in the Hough transform space, and those groupings represent line segments in the edge image. One weakness of global methods like the Hough transform is that they typically group collinear edge segments together even if they are widely separated in the image and therefore unlikely to be related. A better approach is to combine global and local information to group only those collinear segments that are also near each other in the image.

Once line segments have been extracted from an edge image, it is often desirable to locate junctions between them. Junctions are defined as points where two or more line segments meet, and they often represent important features. Junctions can represent corners of objects or points of overlap between objects. Examples of how junctions can be used to help segment overlapped objects are given in Chapters 8 and 9.

6.8 SUMMARY

Edge detection is arguably the most important pre-processing technique. Extracting edges from an image reduces dramatically the amount of data required to represent the image, while retaining much of the information in the original scene. The other pre-processing techniques described above are usually performed in some combination prior to edge detection, to avoid the generation of unwanted or meaningless edges. Some form of edge detection is an important part of nearly every artificial vision system.

The extraction of edge-based primitive features is also important. The first step in processing an edge image is usually the extraction of features like the line segments, vertices and junctions described above. If pre-processing, edge detection and primitive feature extraction have been performed well, the result is an intermediate image representation in which the majority of primitive features are accurately located, are meaningful and represent features of objects of interest present in the original scene.

7

Intermediate image processing

7.1 INTRODUCTION

The techniques discussed in the previous chapter are excellent for eliminating spurious information, clarifying pertinent information, and highlighting important primitive features. The best that can be hoped for from these techniques, however, is the extraction of an accurate edge representation of the scene with important features such as straight lines, corners and overlap points identified. More often the result will be a somewhat inaccurate edge representation, with extraneous features such as shadows, glare and background details still appearing.

In this chapter several intermediate image processing techniques will be discussed. These techniques take advantage of some of the visual clues that humans regularly use to separate objects from the background, to differentiate overlapping objects, or to judge the three-dimensional shape of objects from a two-dimensional image. These techniques use information from color, shading, stereopsis and motion to facilitate the process of eliminating meaningless features and binding meaningful primitive features into groups that represent objects or parts of objects.

7.2 COLOR REPRESENTATION AND PROCESSING

7.2.1 Introduction

The world would not be nearly as interesting to us if it were not in color. Color television sets dominate their market, as do color video cameras, color photographic film, and color computer monitors. Although grey-scale images are used in many computer vision systems, color image representations provide much more information about a scene. As explained in Chapter 5, color can be very useful in many inspection tasks.

Because grey-scale images represent only the brightness in a scene, they are said to convey one-dimensional information. The information conveyed by each pixel in a grey-scale image can be represented by a single number. Because three separate values are required to completely specify a color, color information is said to be three-dimensional. One common color representation is RGB, where the levels of red, green and blue are represented for each pixel. Another color representation scheme uses values that represent hue, saturation and luminance to specify colors. Although these two systems are not identical, they can each represent all possible colors because they are three-dimensional. Representations of one type can also be transformed to representations of another type via simple formulae.

7.2.2 Human perception of color

As was explained in Chapter 3, the human eye contains cone cells that are primarily responsible for color perception. Three types of cone cells have responses to wavelengths that approximate red, green and blue light. The perception of color is not as objective as it may seem, however. The two things that primarily influence the apparent color of an object to an observer are the **reflectance** properties of the surface of the object (i.e. how well the object reflects light at different wavelengths) and the quality of the **illuminant** (i.e. the light falling on the object). The same object can take on many different appearances depending on the color of the illuminant, and most artificial light sources have a wavelength content considerably different to natural daylight. An extreme example is low pressure sodium (LPS) lighting. LPS lighting is sometimes used in locations like parking lots and garages, and it has a yellow hue and is essentially monochromatic. This has the effect of making all of the automobiles in a parking lot look shades of grey until they are driven into more natural lighting conditions. This disorienting condition can be reduced by using artificial light sources that have a wavelength content more closely approximating natural light.

The human visual system compensates well for illuminant effects, but in so doing it becomes susceptible to other phenomena. Consider the illustration in Fig. 7.1, for example. This figure is made up of a number of squares of varying shades, with two squares of interest pointed to by triangular arrows. Which of the two squares appears darker? Most people perceive the square pointed to by the black arrow as being the darker of the two, even though they are exactly the same shade. When the other squares in the pattern are covered, this becomes evident. This phenomenon occurs because one of the squares is surrounded by light neighbors while the other is surrounded by dark areas. Humans perceive this as a change in the illumination level rather than a change in shade, and the perception of the squares changes accordingly.

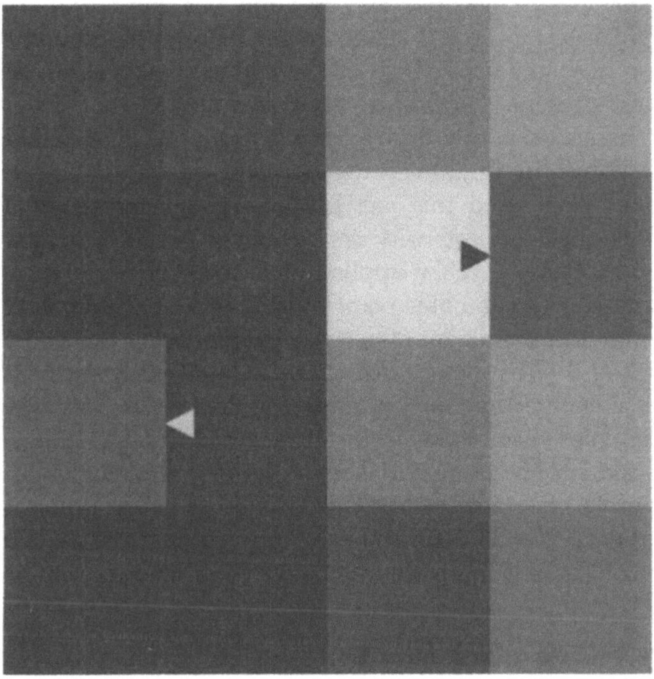

Figure 7.1 Illustration of human compensation for illuminant effects.

7.2.3 Color vision theory

Most of the theories of color vision attempt to explain how humans discount the effects of varying illumination and extract the actual color due to the reflectance of a surface. In one approach, Helson (1938) and Judd (1940) developed a set of equations that explained quite accurately how humans would perceive the color of certain surfaces under certain lighting conditions. This approach has since been expanded and modified by several researchers (Judd, 1960; Pearson, Rubinstein and Spivack, 1969; Richards and Parks, 1971). These equations and their modifications were developed by trial and error, with various terms being added to account for effects observed experimentally. The basic approach is first to determine what the color white would look like under the prevailing illumination conditions, and then to determine the color of any observed area using the computed white reference.

Unfortunately, no underlying explanation for color perception resulted from the Helson and Judd approach. The equations were intended simply to predict the perceived color, not to explain any underlying processing mechanism. Another approach to explaining color vision which was based on underlying physical assumptions, the **retinex**

theory, has been developed by Land and McCann (1971). The retinex theory is based on the fact stated earlier that the perception of color is based on both the reflectance of the surface being observed and the illuminant present. Land and McCann assumed that the illuminant would change smoothly and gradually over the image, whereas any abrupt change in brightness would have to be due to reflectance variations. They used this fact to develop a technique that ignored smooth changes in intensity in an image while emphasizing discontinuous changes. When applied to all three planes of a color image (e.g. the red, green and blue planes) the technique was fairly effective in discounting the illuminant effects while identifying the colors present.

Neither of the theories just described adequately explains all phenomena of color perception. To quote the late David Marr (1982), 'The theory of color vision is in an unsatisfactory and interesting state.' The Helson–Judd approach provides no underlying explanation at all, while some simple cases like sharp-edged shadows cause problems for the retinex theory. If color perception and analysis are to be used in practical situations, it is best to do so under very controlled conditions.

7.2.4 Processing color information

Because the perception of color is so dependent on the illuminant, it is more important than ever to control the illumination used in color vision systems. Illumination sources must be carefully chosen to contain the proper wavelengths, and the effects of ambient illumination must be eliminated to avoid problems. It is not unusual for inspection systems that are susceptible to ambient light effects to give varying results based on the time of day or day of year, or even the presence of local vehicle or pedestrian traffic.

The most effective color processing techniques presently in use take advantage of the three-dimensional nature of color. Because of this nature, any color pixel can be plotted in a three-dimensional color space, with its position in that space indicating its color. It is then possible to find groups of pixels that are in the same region in color space which represent areas of similar color in an image. Often this grouping can be made more effective by transforming the color space (from RGB to hue/saturation/luminance, for example). Pixels only loosely associated in one space may be more tightly grouped in another.

The reverse of the grouping process can also be done, by plotting all of the pixels in color space and then identifying those that fall into a particular region. This is done when a specific color is of interest, as in the case of identifying ripe fruit or when looking for a specific color-coded mark. Only the pixels of the particular color or colors of interest are extracted from the image and analyzed.

Color capabilities make interesting and effective color analysis tools available to artificial vision systems, and represent another step in making machine vision more closely resemble that of humans. As discussed in Chapter 5, however, color vision comes at a cost. Color cameras and processing hardware are usually considerably more expensive than their grey-scale counterparts, and the extra information represented by color images translates into more difficult and time-consuming processing.

7.3 SHAPE FROM SHADING

7.3.1 Introduction

Consider the image in Fig. 7.2. What do you see? Most people perceive this image as a convex or spherical object in front of a featureless background. Now take a moment to turn this book upside down and view the image from that vantage point. Now what do you see? Most people now perceive a concave depression in an otherwise smooth surface. This is in fact a computer generated image that can be interpreted in an infinite number of ways depending on the assumptions that are applied. Humans have strong biases for smooth, continuous surfaces and lighting from above, and those assumptions lead to the common interpretations.

Figure 7.2 A two-dimensional image that can be interpreted as a three-dimensional shape due to shading effects.

The general problem of determining three-dimensional shape by interpreting the intensity levels in a single two-dimensional image is called the shape-from-shading (SFS) problem. Solving the shape-from-shading problem involves reconstructing a three-dimensional surface (usually expressed in terms of height above a reference plane) from the brightness values in the image.

As with many other problems in image processing, the shape-from-shading problem is said to be **ill-posed** because there is not enough information in a single brightness image to unambiguously reconstruct the three-dimensional structure of the objects that appear in the image. The usual approach to eliminating the ambiguities is to apply certain assumptions and constraints as restrictions on the reconstruction process until a single unambiguous interpretation is possible. The first assumption usually involves the types of surfaces that are present in the scene.

7.3.2 Surface models

Two ideal surfaces have commonly been used as approximate models for real surfaces in shape-from-shading problems, and they are the **specular** and **ideal Lambertian** surfaces. Specular surfaces are mirror-like: any incident light ray is reflected in a single direction based on the familiar 'angle of incidence equals angle of reflection' rule. Looking at a specular surface reveals a reflection of the surrounding environment, distorted by the shape of the surface. A polished chrome-plated car bumper is an example of a specular surface.

In contrast to the specular surface, an ideal Lambertian surface is perfectly diffuse. Any incoming light ray is reflected with equal intensity in all directions. Ideal Lambertian surfaces have a diffuse matte appearance. Paper, snow and flat paint are real examples that approximate ideal Lambertian surfaces. It is important to note that the appearance of an ideal Lambertian surface depends only on the location of the light source, not the viewing angle. Because these surfaces reflect incident light equally in all directions, any visible surface patch will not change in appearance as the camera angle is changed.

It is interesting to note that if a perfectly diffuse light source (light coming equally from all directions) is used to illuminate a Lambertian surface, the apparent brightness of all areas of the surface will be equal so no shape will be discernible. This situation is approximated in nature when a snow-covered landscape is viewed under an overcast sky. A very disorienting condition known as **white-out** occurs because surface details and depth can no longer be perceived under these conditions.

7.3.3 Reflectance maps

An intermediate vehicle used to solve shape-from-shading problems is the **reflectance map**. The reflectance map is constructed from the brightness information in the original image and it leads to information about the orientation of the three-dimensional surface that generated the image. The reflectance map can be calculated directly from the brightness values in an image if certain assumptions about the nature of the surfaces represented in the image are applied. If a surface is assumed to be Lambertian, for example, and it is illuminated by a single point source, the reflectance calculation reduces to Lambert's law:

$$R = I \frac{\cos \theta_i}{\pi} \tag{7.1}$$

where $0° \leqslant \theta_i \leqslant 90°$ is the angle of incidence, R is the reflectance, and I is the incident light level.

Figure 7.3 illustrates this situation. Under these assumptions the orientation of any surface patch in an image with respect to the orientation of the light source can be determined directly from the brightness of the patch in an image. Unfortunately, this still does not lead to unambiguous interpretation. Any patch could be rotated about a vector pointing from the patch to the light source, and because the surface is Lambertian the observed brightness of the patch would not change. Put another way, the orientation of a surface patch entails two degrees of freedom, and only one has been eliminated by the reflectance information.

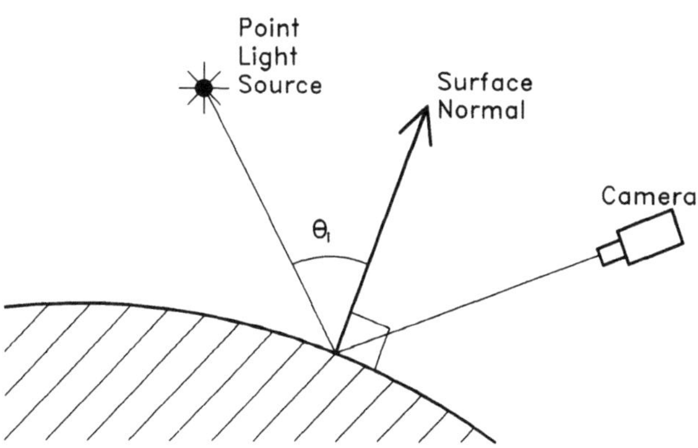

Figure 7.3 Conditions for computing reflectance using Lambert's law.

The remaining ambiguity is usually resolved by making assumptions about the shape of the surface as a whole. Since the reflectance calculation is done locally at small surface patches, the relationships between neighboring patches can be used to determine the actual surface shape. Common assumptions are that the surface is smooth (Ikeuchi and Horn, 1981), or integrable (Frankot and Chellappa, 1990), or at least piece-wise smooth over regions with distinguishable edges between the regions (Malik and Maydan, 1989).

7.3.4 Global shape interpretation techniques

There are two common approaches to applying a global interpretation to the local orientation information and they are energy minimization and propagation. Energy minimization algorithms obtain the solution by minimizing an energy function. Global propagation techniques propagate the shape information from known surface points or seed points to the whole image.

One popular global minimization technique minimizes an energy function consisting of two constraints – the brightness constraint (the recovered surface should produce the same brightness pattern as that observed in the original image) and the smoothness constraint, which forces the gradient of the recovered surface to change smoothly (Ikeuchi and Horn, 1981). Another technique minimizes the same energy function expressed in terms of the surface normal vector rather than the gradient (Brooks and Horn, 1985). Still another utilizes the same energy function but adds the constraint that the surface is integrable (Frankot and Chellappa, 1990), an approach which improved both the accuracy and efficiency of the original algorithm. Finally, in order to provide a solution that would handle piece-wise smooth surfaces, Malik and Maydan (1989) combined both line-drawing and shading constraints in an energy function. Minimizing this energy function they recovered both surface orientation and line locations.

Global propagation techniques typically begin with a singular point, determine the orientation of the area surrounding the point, and then use that information to constrain the orientation of the neighboring surface patches. One technique that uses this approach is the characteristic strip technique (Horn, 1986). A characteristic strip is defined as a line in the image along which the surface depth and orientation can be calculated from reflectance information if these values are known for the initial point of the line. The algorithm starts by identifying singular points (points of maximum intensity are used in this case) as initial points and then determining the orientation of the areas near the singular points assuming a spherical surface model.

This process results in a small 'cap' being formed around each initial point, with the points around the perimeter of the cap having known orientations. Using these perimeter points as starting points, characteristic strips are then constructed in a direction away from each of the initial caps. These strips are then interconnected to reconstruct an approximation of the entire surface shape.

7.3.5 Summary of shape-from-shading

The obvious advantage of applying SFS in any vision system is the ability to reconstruct three-dimensional structure from a two-dimensional image. The resulting three-dimensional structure can then be compared to object models for recognition or inspection purposes.

The disadvantages of shape-from-shading are numerous, however. All of the algorithms require many assumptions to hold, and as such are effective for only certain classes of surfaces viewed under particular conditions. If a situation does not meet the assumed conditions, results usually degrade rapidly. Also, because of their reliance on energy minimization or relaxation techniques, shape-from-shading algorithms are computationally intensive and therefore very time consuming. In inspection systems, shape-from-shading techniques are best used to provide information to assist in the analysis of images rather than to provide actual three-dimensional reconstructions. SFS data can be used to eliminate ambiguities in other processes, for example object segmentation, interpretation of shadows and feature grouping.

7.4 STEREO VISION

7.4.1 Introduction

When looking at the world human beings do not perceive it as a flat, two-dimensional surface. They see a vivid three-dimensional world with depth, and the differences between the real world and two-dimensional representations like photographs, television images, and movies are very noticeable. The most obvious source for this depth information in humans is called **stereopsis**, and it comes from the fact that we have two eyes, each seeing the world from a slightly different viewpoint.

For a simple demonstration of the effects of stereopsis, hold a finger up in front of your face at arm's length against a distant background. If you close one eye at a time you will see the finger 'jump' sideways when you switch between your left and right eye. The relative shift in position of objects in images taken from differing viewpoints is called stereo **disparity**, and the amount of disparity exhibited by an object is an indication of its distance from the viewer.

7.4.2 Stereo disparity and depth

Depth can be computed from stereo disparity in an artificial vision system if two cameras are used to capture images from different viewpoints. Although the problem of computing depth from stereo disparity can be solved in the general case, where the orientation of one of the two cameras is translated and rotated an arbitrary amount with respect to the other (Horn, 1986), a simpler case will be analyzed here for clarity. Specifically, this case involves two identical cameras whose image planes lie in the same plane and whose optical axes are parallel to each other and perpendicular to the image plane (Koenderick and van Doorn, 1976). This situation is illustrated in Fig. 7.4. If the x-dimension is chosen along a line connecting the centers of the image planes for each camera, then any point appearing in both images will appear at the same height (i.e. the same y-dimension) and differ in position only in the x-dimension. In the general case, if a point is found in the left image, due to geometric constraints it must appear somewhere along a straight line in the right image. These lines are called **epipolar** lines, and in the simple case presented here the epipolar lines in the right image are horizontal lines at the same y-dimension as that of the original point in the left image.

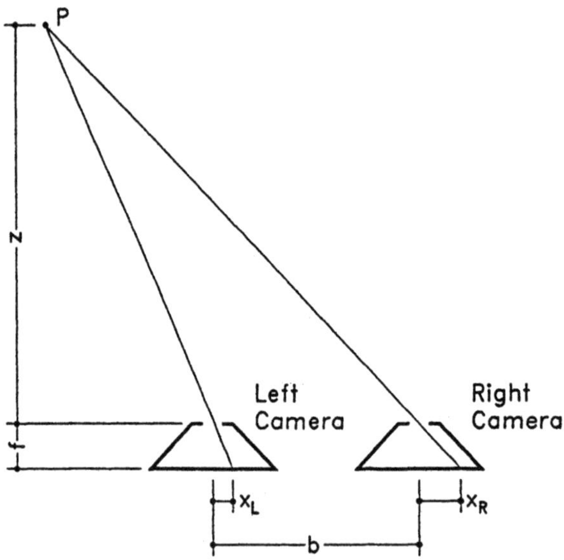

Figure 7.4 Camera geometry for stereo disparity computation.

Based on the geometric arrangement shown in Fig. 7.4, the disparity d is simply calculated as follows:

$$d = x_R - x_L \tag{7.2}$$

where x_L is the x-coordinate of point P in the left image and x_R is the x-coordinate of point P in the right image. Taking advantage of the geometry of similar triangles, it is simple to show that the depth z, or distance from the camera, of point P can be calculated from the disparity as

$$z = f\frac{b}{d} \tag{7.3}$$

where f is the focal length of the cameras and b is the baseline distance between the cameras. Put into words, the distance from the camera is inversely proportional to the measured disparity. This corresponds to actual experience, where relatively near objects seem to 'jump' a greater distance than distant background objects. This fact also has an effect on accuracy. Distant objects exhibit very little disparity so determining their depth accurately is not possible, whereas near objects exhibit larger disparities so their depths can be determined with correspondingly greater accuracy.

7.4.3 Correspondence problem

In the simple situation just explained, the calculation of depth seems almost trivial in its simplicity. By beginning with a point P that was located in both images, however, the explanation neatly dodged the most difficult problem in stereo vision, which is known as the **correspondence problem**. Given a stereo pair of images, any points or objects in the left image must be found in the right image before any calculations can take place. Although the appearance of objects will usually be similar in both images, some distortion is sure to occur due to the differing viewpoints. Also, due to differing fields of view and to near objects overlapping far objects, some points in the left image will not appear in the right image and vice versa. In addition, any object in one image will be in a different position in the other image by definition, and objects of differing depths will change positions relative to each other. All of these facts conspire to make the identification of corresponding points in the two images difficult at best (Allen, 1987).

Some systems have been designed to avoid the correspondence problem entirely. Illuminating the scene with a tiny spot of light from a laser guarantees that only one point will be visible in each of the images, and the disparity and depth can be readily computed. Inspection systems have been designed based on this principle, where many stereo pairs of images are produced as a spot of light is scanned over a scene, and a three-dimensional depth map of the scene is built up based on the depth of each of the points (Marshall and Martin, 1992). This approach has the advantage of eliminating all of the difficulties of the correspondence problem, but it requires that many stereo pairs be analyzed in order to produce a depth map with reasonable resolution.

Most systems employing some form of stereo vision will be faced with having to solve the correspondence problem, and there are some constraints that make this possible. First is the epipolar line constraint mentioned previously, which states that a point in the right image corresponding to a point in the left image must lie along an epipolar line. The location of the epipolar line can be calculated based on the camera geometry and the location of the original point in the left image. In the simple example used previously, epipolar lines were simply horizontal lines at the y-location of the original point. Another constraint states that points of the same depth must occur in the same left-to-right order in the right image as in the left. Another constraint often applied is that disparity must change smoothly over most of the image, with discontinuous changes occurring only where objects overlap.

There have been many approaches to solving the disparity problem (Julesz, 1971; Julesz and Chang, 1976; Nelson, 1975; Dev, 1975; Harai and Fukushima, 1978; Sugie and Suwa, 1977; Marr and Poggio, 1976; Marr, 1980). A rather simple approach will be presented here that has proved itself in practical applications. This approach is based on a technique called **edge matching**, where edges are first extracted from both the left and right images and then the disparity of the edge points is determined. The edge matching technique simply overlays the left and right images on top of each other and shifts them left or right until a peak in the correspondence is found. Remember that, in the simple case of identical parallel cameras, disparity can only occur in the x-dimension. The correspondence peak indicates the correspondence of many edge points in the two images, and all of these points will have similar disparities. The points that correspond at the peak are then assigned the disparity represented by the relative position of the two images, and those points are removed from further consideration. This process is then repeated with the remaining points to determine the next most common disparity represented in the images, and those points are matched and removed. The process continues iteratively until all points are removed or a correspondence peak of sufficient magnitude is not found. Any points not assigned a disparity upon termination are simply discarded. Such points usually represent areas visible in one image but not the other, or inconsistencies in edge detection between the two images.

Figure 7.5 illustrates the results of applying the edge matching algorithm to real images. The image in the figure is of three simple objects (two rectangular and one circular) which are suspended at different depths in the scene. If a simple edge detection technique is used on this image, a single complex object is extracted because the objects all overlap. Once a stereo pair to this image is analyzed, however, and the edge matching process is performed, the objects can be separated based on their relative depth in the scene. The figure demonstrates the effectiveness of the relatively simple edge-matching process for determining areas of common disparity.

This edge matching technique can be performed in parallel by a network similar to a neural network, and as such can be accomplished very quickly. Each node of the network represents each possible disparity, and the winning node represents the peak in correspondence. Once the winning node is determined, the associated edges are assigned the proper disparity value and the output of that node is suppressed in order to allow the matching process to continue. The process continues until no node exceeds a minimum activation level threshold.

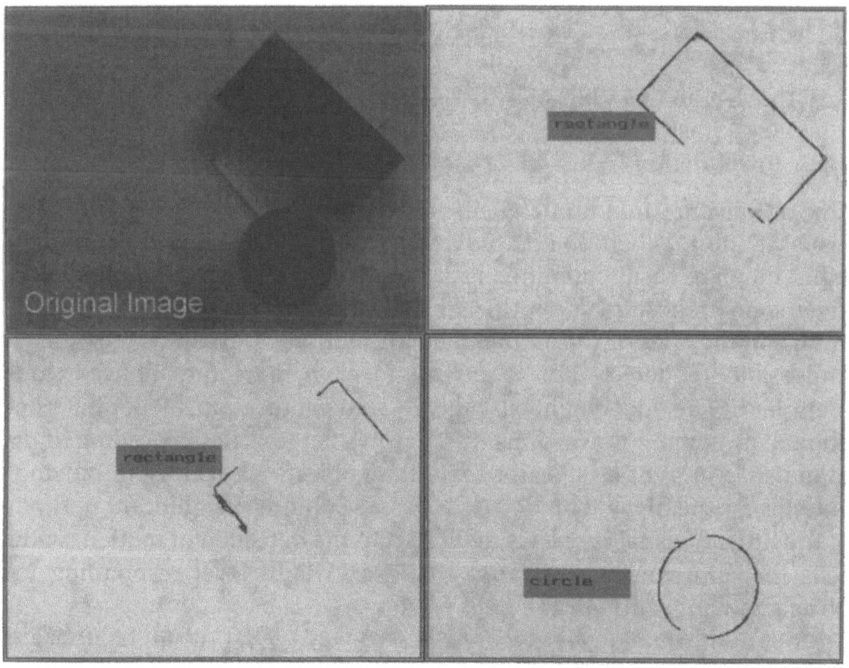

Figure 7.5 Results of using edge matching and stereo disparity to segment overlapping objects.

7.4.4 Stereo vision summary

Incorporating stereo vision into a visual inspection system has the advantage of providing depth information. This information can be used to segment images into regions of common depth quite reliably, as was demonstrated in this chapter. Depth segmentation is often the only way to enable systems to extract and recognize partially overlapping objects from noisy or cluttered scenes. Incorporation of stereo capability could move inspection systems further towards human capabilities in this area.

There are several disadvantages to stereo vision, however. The most obvious is the additional expense. Stereo systems require an additional camera and often extra processing hardware. The additional processing required also increases the time required to perform an inspection. The capability of stereo vision systems to determine the absolute distance of an object from the camera is also limited. Improper camera alignment, limited pixel resolution, lens distortion, and errors in determining correspondence all lead to inaccuracies in the depth computation. For this reason it is more practical to use stereo vision techniques to aid in image segmentation, where relative depth information is adequate, rather than to try to reconstruct an accurate absolute depth map of a scene.

7.5 ANALYSIS OF VISUAL MOTION

7.5.1 Introduction

One of the clues that human beings use to make sense of their world is visual motion. When objects move relative to a background or to each other they are easily identified as separate objects, and the motion also gives some indication of depth. Consider the case of a well-camouflaged animal in the wild. If the animal remains motionless it can be difficult or impossible to detect, but when it moves it is readily noticed and identified. The movement also gives some indication of depth. The animal is coming toward us if it grows larger, it is receding if it diminishes in size, it is nearer to us than objects it occludes in moving and it is farther away than objects it passes behind. A significant portion of the human visual system is dedicated to the detection of motion, with cells that are sensitive to rapid changes in light level responding to objects moving through the field of view.

Motion information can be used in two ways. First, it can be used in the same way that stereo information can be used to segment objects from each other or from the background. Where stereo analysis required objects to differ in depth, motion analysis requires objects to change locations in successive images. Motion analysis can provide information that simplifies the somewhat difficult task of segmenting overlapping objects. The other way in which motion analysis can be used is to provide information on the parameters of the motion itself, in order to determine the speed and direction of movement of an object. Both of these techniques are described in the following sections.

7.5.2 Motion segmentation

The aspect of motion analysis that will be discussed here is called local motion. Local motion is the movement of an object with respect to either the background or another object. Global motion, which will not be discussed here, occurs when almost everything in the scene appears to be moving due to movement of the camera in the environment. The discussion that follows assumes that the camera is stationary.

The typical approach to motion analysis is to analyze the differences between images taken sequentially of the same scene. As stereo analysis requires two images taken from different viewpoints, motion analysis requires a sequence of at least two images taken at different times. Figure 7.6 illustrates this technique on the image of a key taken against the very busy background of a newspaper page. Simply analyzing either of the images in the upper portion of the figure with a computer and trying to extract the key would be very difficult. The general clutter and the fact that the key overlaps similarly colored areas would make this scene a challenge for any analysis technique.

If the key moves against the background, however, the motion can be used to extract the key. The lower image in the figure was generated by subtracting the upper left image from the upper right image and thresholding the result. Pixels that generated a value above the threshold were colored black, while those below the threshold were colored white. Some simple hole-filling could be applied to the result to generate an almost perfect silhouette of the key. If multiple sequential images are analyzed in this way, and the results combined, the method becomes even more reliable at eliminating noise and extracting the true shape of the moving object. In industrial inspection, this kind of analysis could be very useful for extracting objects moving along a conveyor-belt against a cluttered background.

7.5.3 Determining motion parameters

Although the segmentation technique just described is very useful, it is often desirable to extract information about the motion itself in addition to the information about the shape of an object. A technique for determining the speed of a moving object is described next, and slightly

Figure 7.6 Taking advantage of the presence of visual motion between two images (a) and (b) to extract a moving object (c) from a cluttered background.

more sophisticated techniques can be used to extract the direction of the motion as well.

In addition to the stationary camera assumption applied previously, extracting speed from motion requires that even more restrictions be placed on the situation being analyzed. First, the motion must be simple linear translation (no rotation) and it must be perpendicular to the camera line-of-sight. The motion is also required to be oriented with the camera so that translation occurs only in the camera x-dimension, and the object must be far enough away from the camera so that the apparent size of the object does not change appreciably in successive images. The distance from the object must also be known together with the geometry of the camera, so that distances measured in camera pixels can be translated into real distances in the environment. Finally, the time between images must be accurately known if speed information is to be determined with precision. Although these constraints may seem overly restrictive, they are usually met by the situation of an object moving on a conveyor-belt, with the camera aligned to the conveyor and images being acquired at regular intervals.

Under these conditions the approach used to determine speed is relatively simple. First, the object is located in two successive images, and the number of pixels that the object has moved between the images is measured. The geometric information and the time between images are then used to compute the actual distance and speed of movement.

The phrase 'the object is located in two successive images' should have a familiar ring to it, as it is another instance of the correspondence problem. Because the translation has been limited to only the camera x-dimension, the edge matching technique for determining correspondence described in section 7.4.3. can be used again with reasonable results. The problem is even simpler, however, if at least three successive images are available. The subtraction technique described in the previous section for segmentation can then be used for the 1–2 image pair and the 2–3 image pair, and the results can be compared. The translation of the silhouette between the two resulting images represents the movement of the object. Applying edge detection and edge matching to these silhouettes usually results in a very reliable determination of correspondence, and an accurate translation measurement. The actual speed of the object can then be computed, based on the camera geometry and the time between images.

7.5.4 Summary of motion analysis

First, it must be emphasized that the two types of analysis described above involve very simple cases. Motion analysis becomes far more difficult if objects are moving towards or away from the camera, are rotating, are non-rigid (changing shape), or some combination of these. Far more sophisticated techniques are required to derive motion information in these more difficult cases.

With that said, even the simple cases presented here have limitations. Because the objects of interest must be present in successive images, the speed of movement must be low enough so that they do not pass out of the field of view. There is a lower limit to the accuracy of speed measurement as well, since the measurement relies on distances measured in camera pixels. Movement of less than 2 or 3 pixels in successive frames will result in very poor accuracy in the speed measurement. Other weaknesses are similar to those noted for stereo analysis, since the computation of motion is also reliant on accurate camera geometry and a reliable solution of the correspondence problem. For these reasons motion analysis, like stereo analysis, is best used as an aid in image segmentation. As was demonstrated in Fig. 7.6, taking advantage of motion information can transform the very difficult problem of extracting an object from a cluttered background into a relatively easy problem of silhouette analysis.

7.6 GROUPING PRIMITIVE FEATURES INTO COMPLEX FEATURES

Even without additional information provided by the techniques just described, a simple intermediate edge image with primitive features identified allows some geometric grouping techniques to be applied. Proximity can be used to group features, since features that are nearby in an image are often associated with the same object. Collinearity is another grouping aid, since linear edge segments that are aligned often belong to the same object. Containment is another geometric grouping technique. Features contained within other features often belong to the same object, as in the case of printing on the surface of an object. The printing may not touch the outline of the object, but it is contained within that outline.

The intermediate processing techniques presented in this chapter allow even more powerful grouping methods to be used. Areas of common color often belong to the same object and can easily be grouped if color capabilities are present. If shape analysis is used, features that represent a smooth, continuous three-dimensional shape can be grouped together. Depth information provided by stereo vision can also be used to group features of common depth. Finally, features that move together often belong together and can be grouped by motion analysis. It was repeatedly stated above, however, that all of these methods are imperfect to some degree, so none of them provide enough information by themselves to perform reliable grouping and segmentation. Combining information from two or more of these techniques, however, can lead to the reliable extraction of object representations from images.

7.7 SUMMARY

First, it should be said that what has been provided here is a very brief overview of some intermediate image processing techniques. The intention was to provide the reader with an overall understanding of what is possible in the way of intermediate processing, and not to explain each technique in detail. Entire books have been written on each of these topics, and much additional information can be derived from the references cited here.

The techniques explained here can be very useful for grouping primitive features into representations of individual objects. Consider the inspection of printed beverage cans moving down a conveyor. Common movement can be used to extract the cans from the background, proximity to group the features that belong to each individual can, and containment to associate printed information with the can that it is on. Shape analysis could reveal any deformities in the can or printing, and color analysis could determine if the base color of the can and the printing color were correct. In short, the intermediate processing techniques enable a far more complete and intelligent inspection than that provided by the analysis of edge information alone.

8

Computational approach to artificial vision

8.1 INTRODUCTION

The effort to model the intelligent behavior of humans with mathematical formulae and computer algorithms is as old as computer science itself. Ever since Alan Turing tried to define artificial intelligence in the 1950s (Hodges, 1983), researchers have sought to create it. Early in these efforts it became obvious that the processing of sensory inputs was a primary source of human intelligence, and that contemporary computers were ill-suited to the task of processing the huge amounts of sensory information available to the human brain. The human visual system is a prime example of an instance of this problem. Even though the practical implementation of artificial vision systems was far off, however, researchers worked diligently to develop the theories, equations and algorithms that would imitate and/or explain the behavior of the human vision system. Steady increases in computer power have made these efforts more realistic, and have spawned a corresponding increase of activity in this area. The result of all of this activity has been the development of several computational models of human vision.

In this chapter the general structure of computational models of vision are explained, and examples of such models are described. The coverage begins with the work of the late David Marr and his associates, who developed prior to 1982 such a computational model. Although this model is presently not state-of-the-art, it represents the first complete attempt at modeling the human vision system, and it introduced many concepts that became important parts of subsequent research efforts in this area.

8.2 THE WORK OF DAVID MARR

8.2.1 Background

Marr described vision as a process that produces from images of the external world a description that is useful to the viewer and not cluttered by irrelevant information (Marr, 1979). Much of the model that was the result of the work of Marr and his associates is based on the biological evidence presented in Chapter 3. The model is broken into three stages, each of which performs a certain task, and all of which must be executed in series (i.e. the second stage requires as input the outputs supplied by the first stage, and so on). Since the specifications of these three stages have become part of the vocabulary of artificial vision, they will be described in a general context as well as in the specific context of the Marr model.

8.2.2 First stage: early vision

The result of early vision, according to the Marr model, is the **primal sketch** (Marr and Poggio, 1976). The first process in creating the primal sketch is edge detection. An edge is defined as a point which coincides with a spatial discontinuity in intensity in the image. The technique used to calculate edge values in the Marr model is the Laplacian-of-Gaussian (LoG) technique discussed in Chapter 6. The LoG operator was chosen because it is a close approximation to the processing thought to occur in biological vision systems. This faithfulness to the biological evidence is typical of the Marr model.

Once the LoG edge detection operator is applied to an image in the Marr model, each pixel is replaced by its corresponding edge value. In order to complete the primal sketch, the stronger of these edge points must be grouped into line segments representing the actual edges present in the original image. The Marr model takes advantage of local geometry in order to group these edge points into line segments. First, significant points are connected by virtual line segments, representing candidate edges. Then some rules are applied to these virtual line segments. In particular, short segments are favored over long segments, pairs of parallel segments are favored, and collinear segments are favored. In this way some segments are eliminated and some chosen as actual edges. Each chosen segment is assigned a value indicating the strength of the edge (reflecting the strength of the local gradient in the original image), and each has a representative orientation. The primal sketch, then, is the two-dimensional collection of these edge segments. Very much of the original image information is preserved in the primal sketch representation, and Marr showed that the original image can be reconstructed quite well by working backwards from the primal sketch (Marr, 1979).

8.2.3 Second stage: intermediate vision

In the intermediate stage, the Marr model seeks to construct a representation of the image which indicates the surface orientation and depth, with respect to the viewer, of each region in the image (Marr and Nishihara, 1977). This reconstruction is an ill-posed problem: from a given primal sketch many different constructions are possible. With all such problems some technique must be devised by which to decide between available alternatives. Typically either additional sources of information are brought to bear which remove the ambiguity, or constraints are placed on the possible solutions which have the same effect. The Marr model takes the first approach, utilizing information from the image in the form of texture gradients, shading, stereo disparity and motion analysis (Chapter 7) to help determine the shape and depth of the represented surfaces. This bottom-up approach is an important aspect of the Marr model.

Marr argued that images contain virtually all of the information required for their interpretation (Marr, 1982), whereas other researchers have taken the position that people rely on *a priori* knowledge of the world in the form of known objects or relationships in order to constrain the solution of the interpretation problem. This latter approach is known as top-down processing, and systems discussed later in this chapter will be seen to rely heavily upon it.

In order to construct the intermediate representation, then, the Marr model assumes some simple surface primitive (a small rectangular element is usually adequate), and then calculates the orientation and relative depth of each of the surfaces represented in the primal sketch based on the sources of information previously described. From this representation the location of surface discontinuities are calculated, and missing edges in the primal sketch are replaced. This intermediate representation is known as **the 2½-D sketch** (Marr and Nishihara, 1977), and it is usually presented as a **needle diagram**. In a needle diagram, the orientation of each surface is represented by a distribution of fixed length vectors (or needles) perpendicular to the surface orientation at the point of attachment (Horn, 1977). The vectors are drawn to appear three-dimensional, that is vectors parallel to the line of sight are drawn as points, and vectors perpendicular to the line of sight are drawn at their full length, while those at other angles to the line of sight are drawn with appropriate lengths. An example of the 2½-D sketch for a cube is shown in Fig. 8.1.

The 2½-D sketch exhibits two important aspects. First, it is constructed

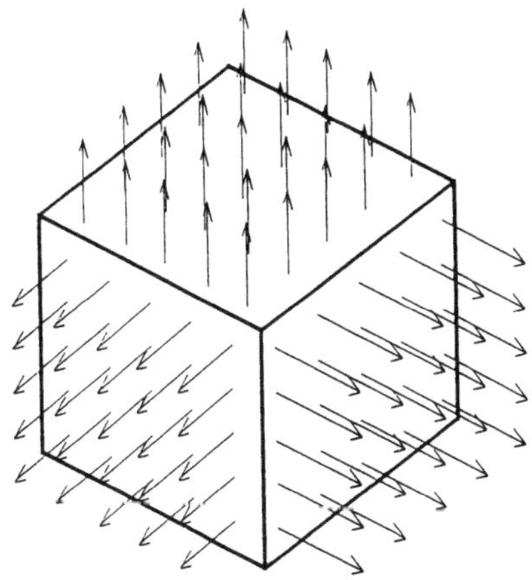

Figure 8.1 The 2½-D sketch of a cube.

in viewer-centered coordinates. This means that the same object viewed from different viewpoints will yield different 2½-D sketches. The representation is not viewpoint invariant. Also, no information beyond what is present in the original image is represented. Only the visible surfaces are shown, and hidden structure is not assumed or guessed at. This is why it is called the 2½-D sketch, because it is not a true 3-D representation of objects.

8.2.4 Third stage: higher vision

Higher vision deals with the identification of familiar objects from visual information. In the case of the Marr model, a transition must be made from the 2½-D sketch representation to some internal representation of known objects. Two primary problems are encountered when trying to accomplish this. First, a general representation must be found which can be used to describe virtually any three-dimensional object. Next, these objects must be described in object-centered coordinates, so that they will be recognized irrespective of scale, rotation and position.

The Marr model solves the first problem by introducing the concept of **generalized cones** (Marr and Nishihara, 1977). A generalized cone (GC) is the three-dimensional shape that is generated when a planar cross-section of fixed shape, and smoothly varying size, is moved through space along an axis. Some example GCs are shown in Fig. 8.2. Marr argued that most shapes, particularly man-made shapes and natural shapes that are the result of a growth process, can be represented by connected groups of GCs. This representation also conveniently solves the other problem: object coordinates are centered on the axis of the

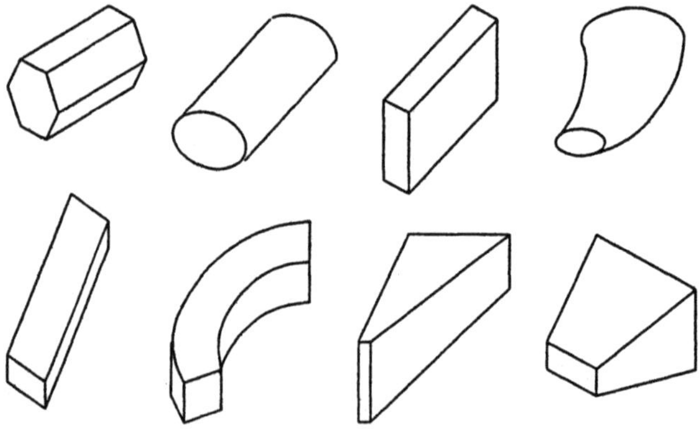

Figure 8.2 Some example generalized cones.

generalized cone. If the cross-section exhibits some symmetry, that fact can be used to further orient the coordinate system.

The Marr model breaks objects down into a hierarchy of generalized cones, with more detail available at lower levels. For example, a man may be represented as a single GC with an elliptic cross-section (resembling a mummy) at the highest level, while at lower levels the major parts of the body would each be represented by GCs, and at still lower levels each of those parts would be broken down into more GCs and so on (Marr, 1982). This hierarchical representation obviously lends itself first to course and later to fine feature matching, another trait supported by biological evidence.

The problem of generating generalized cone representations from the 2½-D sketch was still ill-posed, however, so Marr introduced the concept of **planar contour generators** (PCGs) as a constraint (Marr, 1977). This constraint simply states that the contours viewed in the image are assumed to be generated by a plane, perpendicular to the line of sight, that intersects the surface of the object. Although this constraint is not strictly in agreement with human performance, there is some psychological evidence for the presence of a similar constraint in humans. Once the PCG constraint was assumed, GCs could be generated from the 2½-D sketch, and groups of GCs could then be compared to known models in order to complete the recognition process. An overall diagram of the three stages of the Marr model is shown in Fig. 8.3.

8.2.5 Summary of the Marr model

The Marr model has some drawbacks. First, although some modules of the model have been realized, the entire model has never been implemented and tested on real hardware. Second, the sequential nature of the model (Fig. 8.3) would suggest that, if humans were

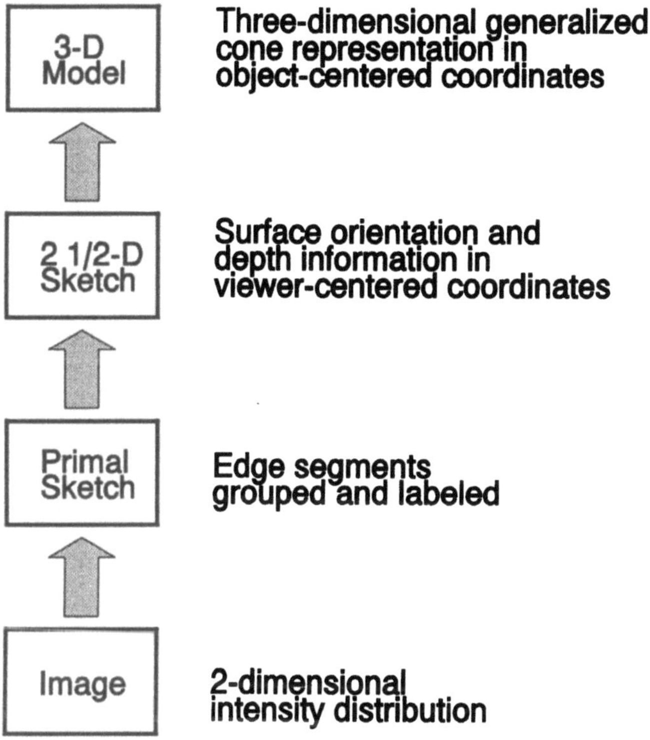

Figure 8.3 Processing stages of the Marr model.

exposed to brief visual stimuli, they would report first a primal sketch perception, then a 2½-D sketch and finally recognition. No evidence exists to support this (Haber. 1983), and in fact the stages of perception have been found to be quite different (Chapter 3). Finally, several classes of objects cannot be adequately represented by generalized cones, and therefore would be impossible to represent or recognize in this system. Even with these drawbacks, however, the Marr model has become the cornerstone of work on computational models for vision. It represents the first complete and rigorous attempt to model human visual performance, and it provided a starting point for virtually all of the models that followed it.

8.3 ACRONYM VISION SYSTEM

8.3.1 Overview

The ACRONYM system, developed by Brooks (1981), is a model-based system of visual object recognition. This system relies heavily on *a priori* data from known models, and it therefore has a strong top-down flavor as opposed to the bottom-up approach favored by Marr. The model data must be entered into the system manually. ACRONYM uses an iterative matching technique between model features and image features to identify objects in the image.

8.3.2 Input

ACRONYM takes as input an edge-detected image. The line finding technique of Nevatia and Babu (1980) is used to extract the edges, and the edges represent all of the information extracted from the image. None of the other sources of information available (shading, stereo disparity, motion, etc.) are utilized. This simplicity is both a strength and a weakness. Limiting the amount of image processing obviously makes the model efficient, but inevitable errors in edge detection can propagate through to the recognition phase, and no other sources of information are available to eliminate ambiguities.

8.3.3 Models

ACRONYM uses parameterized three-dimensional models, with model-based coordinates. The models consist of primitive volume elements and their relationships. The volume elements are generalized cones, as introduced by Marr. The GCs are specified by a planar cross-sectional shape, an axis or spine along which the cross-section is swept to generate the shape, and a sweeping rule which governs how the size of the cross-section changes along the spine. More complex objects are represented as interconnected graphs of GC elements, with the GCs comprising the nodes of the graphs, and the arcs of the graphs representing the type and orientation of the attachment between the connected GCs. These graphs are organized in a hierarchy based on level of detail, with the gross shape of the object residing at the top and fine details residing at the bottom. A representation of such a hierarchy for the model of an electric motor is shown in Fig. 8.4. As in the case of the Marr model, coordinate systems for each GC element are centered on the axis.

Attachments between GC elements are represented as rigid or articulated. The representation for rigid attachments includes information on the relative coordinate transformation which takes place between the two connected GCs at the attachment point, so complex objects can be represented accurately. The representation for articulated attachments indicates the axes of articulation as well, so connected GC elements can be allowed to move with respect to each other.

None of the model specifications, or parameters, need to be fixed. The

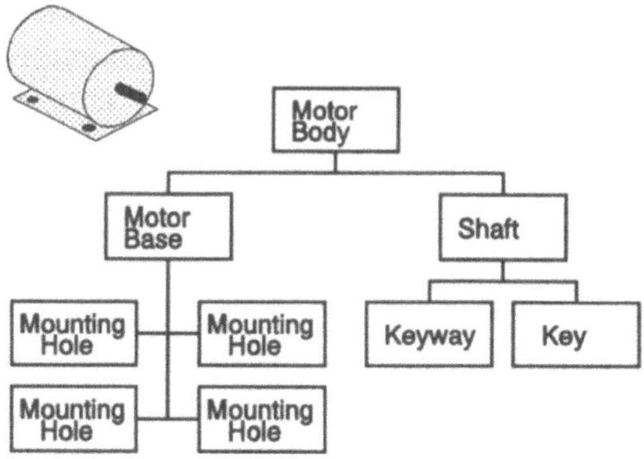

Figure 8.4 Example graph showing interconnections between components of an electric motor.

parameters may vary within certain constraints (numerical inequalities), and the constraints can be determined by algebraic expressions involving the values of other parameters. In this manner general classes of objects can be represented, while fixing the parameters at certain values represents an instance of the class. This ability to generalize as well as to make specific decisions about objects is an important aspect of ACRONYM. For example, while searching an aerial photograph (Brooks, 1981), ACRONYM could determine how many total airplanes were in the photograph (general), as well as determining if a Boeing 747 were present (specific).

8.3.4 Geometric reasoning

The ACRONYM system works by predicting what features will be present in an image, and then searching for those features. For this reason a geometric reasoning system was developed to determine what features of a given model would be visible from a certain viewpoint, and what relationship those features would be in. Viewpoint invariant features, like parallel and collinear lines, are also identified. The geometric reasoning system uses a combination of spatial mathematics, coordinate transformations, geometry and algebra to predict how the three-dimensional model will appear from a certain viewpoint, and what features or relationships will be evident from nearly any viewpoint.

8.3.5 Model matching and image interpretation

In order to match three-dimensional models to two-dimensional images, a translation of some type is required. ACRONYM solves this problem by introducing the concepts of **ribbons** and **ellipses** (Brooks, 1980). Ribbons are planar representations of the silhouettes of generalized cones. Ribbons are characterized by having a line segment as a cross-section, a spine or axis which is a line or smooth curve in a plane and along which the cross-section is swept, and a sweeping rule which determines how the length of the cross-section changes along the axis. Ribbons are obviously 2-D versions of GCs. Ellipses simply represent the visible ends of generalized cones. The features generated by the aforementioned geometric reasoning system are specific ribbons and ellipses in certain relationships with each other.

Once target ribbons and ellipses are determined, the image is searched for any instances of them. A graph is constructed to represent the image, with nodes representing the ribbon and ellipse features and with arcs representing the relationships (specifically connectivity) between the features. The connected graph of ribbons and ellipses from the image is then matched to the graph generated by the model-based prediction process. If a rough match is found, the parameters implied by the image ribbons are sent back to the geometric reasoning system, and a finer prediction of appearance is generated. This process is repeated iteratively until either an exact match is found, or the matching process diverges. In the aforementioned airplane identification example, predicted ribbons representing wing and fuselage shapes were searched for in the image. Several matches for each were found, but most were improperly connected and were therefore eliminated. Parameters such as wing sweep angle were then computed from the remaining image ribbons, and a more precise prediction was generated which included additional features to search for. In this manner a search was first conducted for the fuselage and wings, and then for tail pieces, and further for window openings or specific markings that would identify a certain aircraft. With each iteration the parameters were further constrained, and the identification became more precise.

8.3.6 ACRONYM summary

The ACRONYM system employs very simple image processing, working directly with edge-detected images. The system has a strong top-down component, and as such relies heavily on model data input manually. The system is not capable of acquiring knowledge of models directly from images. ACRONYM uses geometric reasoning about models to predict which features will be present in an image, and it refines those predictions based on information derived from the image in subsequent iterations. ACRONYM has been implemented in a computer system, but specific performance information has not been published. Geometric reasoning, iterative prediction and graph matching and searching routines, however, do not normally lend themselves to efficient implementation.

8.4 SCERPO VISION SYSTEM

8.4.1 Overview

The SCERPO system for identifying objects in images was created by Lowe (1985). SCERPO stands for **spatial correspondence, evidential reasoning, and perceptual organization**, and the system is an extension of previous research in the vision area, primarily that of Marr and Brooks presented previously. The approach employed in SCERPO is similar to that used in ACRONYM, in that a top-down analysis of spatial correspondence between known models and image features is conducted. The primary difference between the two approaches is that SCERPO uses perceptual organization concepts to group features and to perform spatial inferences at the intermediate stage.

8.4.2 Importance of line drawings

Lowe argues that the fact that objects can be recognized by humans from individual two-dimensional line drawings is critical to artificial vision research. There are many examples of this capability in everyday life, but a definitive experiment to verify this capability in humans was conducted by Hochberg and Brooks (1962). In this experiment a child was raised to the age of 19 months without being exposed to two-dimensional drawings of any kind. All of the child's playthings were made of solid colors, and even two-dimensional markings or patterns were eliminated from the environment. The child was in no way allowed to learn to associate drawings of objects with real objects.

After this 19-month period, the child was asked to identify familiar objects (a car, a truck, a key, a shoe, etc.) from simple black-on-white line drawings. These drawings were constructed without regard to scale, rotation, or viewpoint. For example, the car, the key, and the shoe were all drawn approximately the same size. The child had no difficulty identifying any of the objects in the drawings. Later tests with black-and-white photographs yielded identical results: the child recognized all of the objects.

These results suggest that humans have the built-in ability to recognize line drawings, or that this ability is inherently developed along with the ability to recognize three-dimensional objects. The ability to associate between line drawings and the objects that they represent does not have to be learned or developed. Lowe argues that, in these experiments, the child had no opportunity to form a 2½-D sketch representation of the drawn object and recognition proceeded uninhibited anyway. For this reason SCERPO works with only line drawings derived from edge-detected images, and does not derive information from depth, shading, stereo disparity or motion. The edge-detection technique is a modified version of that used by Marr, and it applies the LoG operator. Once edges are detected, significant line segments are extracted from the resulting image, and these are the raw features that SCERPO works with in the recognition process.

8.4.3 Perceptual organization

The centerpiece of the approach used in SCERPO is the application of the principles of perceptual organization. These principles were first formalized by the Gestalt psychologists, as explained in Chapter 3. SCERPO employs perceptual organization rules to group primitive features into more complex objects. Lowe demonstrated this tendency in humans by presenting subjects with a drawing of a bicycle that was made up of short unconnected line segments (Fig. 8.5). The drawing was in no way complete, and the segments used were specifically chosen to discourage perceptual organization. When shown this drawing, the majority of the subjects could not identify it in 60 seconds. A single line segment was added to this drawing in such a way as to make more evident, by grouping nearby segments, the outline of the front wheel (Fig. 8.6). When subjects were presented with this new

Figure 8.5 A fragmented line-drawing to test perception. (Redrawn from D.G. Lowe, *Percéptual Organization and Visual Recognition*, Kluwer Academic Publishers, Norwell, MA, 1985, Fig. 1-4. Copyright © 1985 Kluwer Academic Publishers. Used by permission.

Figure 8.6 The drawing of Fig. 8.5 with a single segment added. (Redrawn from D.G. Lowe, *Perceptual Organization and Visual Recognition*, Kluwer Academic Publishers, Norwell, MA, 1985, Fig. 1-5. Copyright © 1985 Kluwer Academic Publishers. Used by permission.

drawing, the majority of them easily recognized the bicycle in less than 60 seconds, some doing so in 5 seconds or less. This experiment provides evidence of the strong role that perceptual organization plays in feature grouping and object recognition.

Several perceptual organization concepts can be used for locally grouping line segments: collinearity, parallelism, proximity of endpoints and symmetry are some examples. When simple features can be grouped into more complex features, the search for possible object matches can be considerably constrained. Without such grouping, all possible combinations of features present must be considered in the recognition process.

Perceptual organization can also provide a three-dimensional inference capability. Viewpoint-invariant groupings such as parallelism and collinearity can provide three-dimensional information from two-dimensional images. The basic rules applied by SCERPO are as follows:

1. *Collinearity* Three or more line segments that are collinear in an image infer collinearity in three-dimensional space.
2. *Parallelism* Line segments that are parallel in an image are also parallel in three-dimensional space.
3. *Terminations* Two or more line segments terminating at a single point in an image also terminate at a single point in three-dimensional space.

Exceptions to each of these rules can be contrived, but Lowe argues that the probability of accidental occurrence is small and, when several of these rules lead to the same conclusion, that the probability of the result being accurate is very high. The goal of the application of these rules is to provide a partial segmentation of the image into sets of related features, and to generate constraints on the three-dimensional relationships between components of a scene.

8.4.4 Model-based matching

Like the ACRONYM system, SCERPO proceeds by searching for a viewpoint of a model that would result in the observed projection of features in an image. SCERPO models are rigidly defined three-dimensional constructs, but they could be parameterized like those of ACRONYM. Model data is manually entered, and used in a top-down manner. The grouping of features by the previously stated rules of perceptual organization is explicit in the models, i.e. groups of collinear, parallel and co-terminating features are indicated when the model data is entered.

SCERPO begins the recognition process by first grouping the line segments extracted from the image in the manner previously described. These groupings are then compared with those of the models, and a model is chosen and a rough viewpoint is guessed at. Using this viewpoint, the chosen model's features are then projected into the image plane, and the degree of match is calculated. Based on the results of the matching process, a new viewpoint is calculated and the procedure is repeated. This process proceeds iteratively until either it diverges, or a match is found between the model features and image features.

8.4.5 SCERPO system implementation

The SCERPO system was completely implemented on computing hardware. The edge detection was performed by a special purpose vision processor, and the detected edge points were then transmitted to a VAX computer. This computer searched for the significant line segments in the edge image, and performed the perceptual organization groupings based on parallelism, collinearity, and endpoint proximity of the line segments. These groupings were performed only in a limited region around each feature, so proximity of line segments was an inherent prerequisite for grouping. Once grouping was complete, the recognition process previously described was performed. The degree of match between projected model features and actual image features was computed using a least-squares technique. The recognition process typically converged to minimum error in only four iterations. A block diagram of the entire image analysis process employed by SCERPO can be seen in Fig. 8.7.

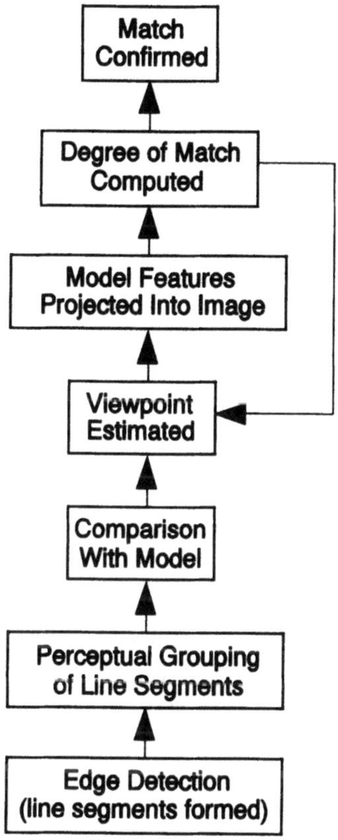

Figure 8.7 SCERPO image analysis procedure.

The following elapsed times (seconds) were reported as typical values for the processing done in the SCERPO implementation:

Edge detection	30
Line segment extraction	40
Indexing and grouping	60
Matching and verification	40

These times were recorded during trials on relatively simple images, and where very few models were used. In fact, some trials involved a search for a single model in an uncluttered image containing only one instance of the object.

8.4.6 SCERPO summary

Like the ACRONYM system, SCERPO works directly with edge-detected images, has a strong top-down component, and relies heavily on model data input manually. The primary contribution made in the development of SCERPO was the introduction of the idea of using perceptual organization concepts to group features and to conduct three-dimensional inference. This capability allowed SCERPO to significantly reduce the search space of possible matches and possible viewpoints, and it duplicates a human tendency. The technique employed by SCERPO to compute new viewpoints from match data performed very well, requiring only an average of four iterations to converge within the limits of accuracy of the data. The SCERPO system has been implemented in a computer system and encouraging results were obtained, although they were based on the analysis of relatively simple images with very few models.

8.5 RECOGNITION BY COMPONENTS THEORY AND THE PARVO VISION SYSTEM

8.5.1 Introduction

A theory that explains many of the aspects of human visual perception of objects has been presented by Biederman (1985). This theory, called **recognition by components** (RBC), is based on the assumption that objects, like words, can be represented by connected combinations of a relatively small number of primitive components. Speech involves the combination of sounds into words. In the English language, virtually all spoken words can be constructed from a set of 38 sound primitives called **phonemes**. The Hawaiian language consists of only 15 phonemes, and the combination of all human languages spoken on Earth can be represented by 55 phonemes. Biederman argues that the system for understanding images operates in a manner similar to that for understanding words, that is that objects are separated or parsed by some process into a connected group of primitive components. These components, furthermore, are such that they can be recognized from simple two-dimensional line drawings, without the use of depth or surface information. Surface characteristics such as color or texture, according to Biederman, are used to make fine discriminations among objects of a single class at a higher level of processing.

8.5.2 Recognition by components theory

Biederman defines a set of five relationships between image features that he calls the non-accidental relations. Each of these, when present in a two-dimensional image, is assumed to imply a similar three-dimensional relationship. The relationships are illustrated in Fig. 8.8. The collinearity relation states that if line segments are collinear in a two-

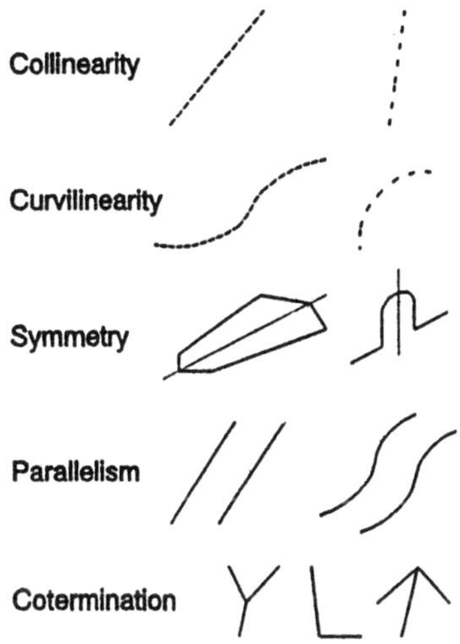

Figure 8.8 The non-accidental relationships.

dimensional image, then they are also collinear in three-dimensional space. Curvilinearity states that line or curve segments that form a smooth curve in a two-dimensional image also form a curve in three-dimensional space. Symmetry states that symmetric objects in an image (the symmetry may be skewed due to viewpoint) are also symmetric in three-dimensional space. The parallelism relationship states that two-dimensional parallelism implies three-dimensional parallelism, and the cotermination relationship holds that line segments that terminate at a common point in a two-dimensional image also terminate at a common point in three-dimensional space.

These relationships are obviously developed from the Gestalt laws of perceptual organization explained in Chapter 3, and are nearly identical to the perceptual grouping and three-dimensional inference rules applied by Lowe in SCERPO. Biederman argues that the accidental occurrence of these relationships in a two-dimensional image is highly unlikely, and furthermore when such accidents occur humans are likely to make a perceptual error and assume that the relationships hold anyway (Ittleson, 1952). Because the non-accidental relationships are invariant from three-dimensional space to two-dimensional projections, they play an important role in the RBC theory. In particular they provide the primary means of discrimination between object primitives.

Object components

As previously stated, the RBC theory rests on the fact that objects can be divided up into simple components. Like Marr, Biederman suggests the use of generalized cones to represent the primitive volumetric components. The way in which the non-accidental relationships differ over the range of generalized cones is used to divide the cones into different classes, and each of these classes represents a different volumetric component as defined by RBC. GCs are classified as to whether the cross-section is made of straight or curved lines, whether the cross-section is rotationally symmetric, reflectively symmetric, or asymmetric, whether the cross-section is constant in size, or expands along the axis, or expands and contracts along the axis, and whether the axis is straight or curved. This representation leads to the definition of 36 different volumetric components, as can be seen in Fig. 8.9.

Since surface features on objects often exhibit no depth, planar primitives must be represented as well. Biederman applies the same

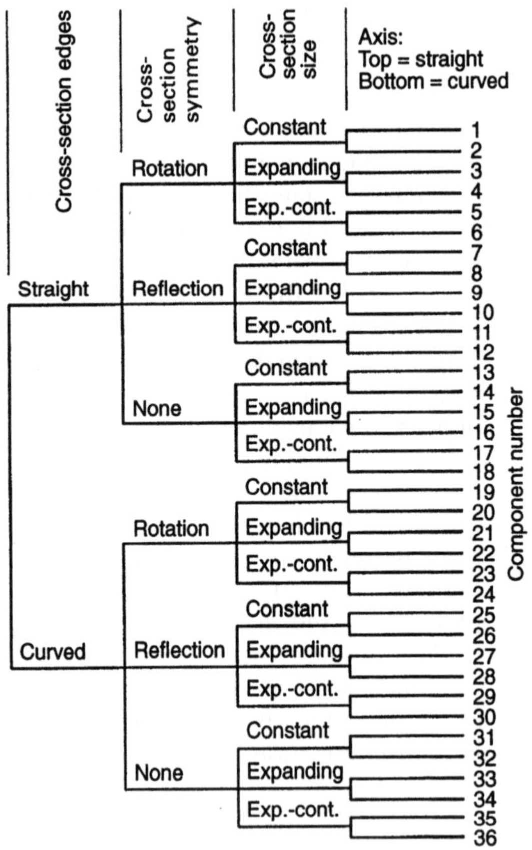

Figure 8.9 Volumetric components used in the recognition by components theory.

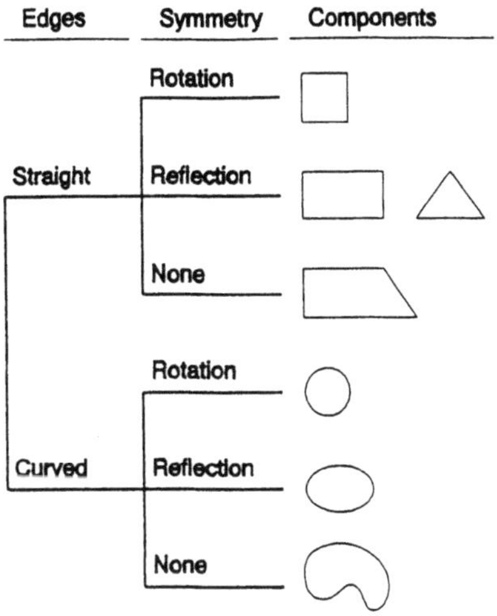

Figure 8.10 Planar components used in the recognition by components theory.

definitions of symmetry and straight/curved edges to planar objects as he does to cross-sections of GCs, generating seven more components as shown in Fig. 8.10.

Some may argue whether 43 components are enough to represent all known objects, but Biederman presents some convincing arguments. First, the number is similar to the number of phonemes, and those are enough to represent all known words. Surely the set of names of known objects is a subset of that set. Also, phonemes can only be connected one-dimensionally, i.e. they can precede or follow each other, whereas primitive object components can be connected in three-dimensional space, allowing a much larger number of possible combinations. Finally, Biederman has shown that the number of possible three-dimensional connections of a relatively small number (less than 10) of the 36 volumetric components numbers in the millions. This comfortably represents the number of objects that can be learned in a lifetime.

Object subdivision

The RBC theory states that objects are separated by the visual system into components, and that these components are then each represented as one of the defined primitives. The primary mechanism for dividing objects into components is separation at cusps, or points of extreme concavity (Hoffman and Richards, 1985). Figure 8.11 illustrates how three common objects, a cup, a table and a flashlight, can be separated into components in such a manner. This method of subdivision emphasizes the importance of T-junctions in two-dimensional line

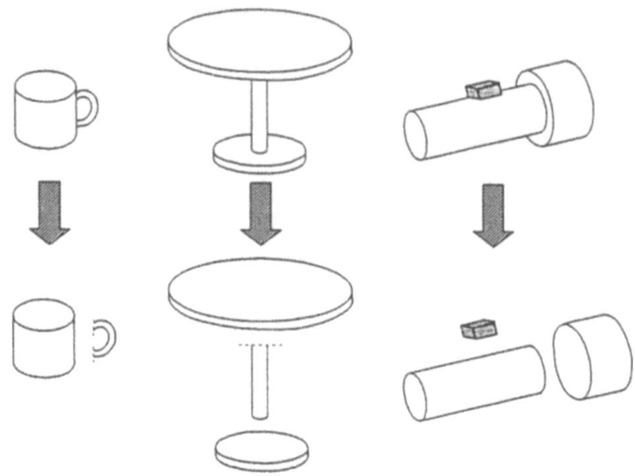

Figure 8.11 Example showing segmentation of common objects into components.

drawings. A T-junction occurs when a line segment terminates somewhere on another line segment, forming the T-shape. T-junctions are almost always formed where one component of an object is physically joined to another object (as between the handle and body of the cup in Fig. 8.11), or when one component occludes another (as with the top of the table and the table support). T-junctions are not considered coterminations, but rather are used to imply component separation in three-dimensional space.

Summary of the RBC theory

Biederman has conducted a comprehensive series of experiments in order to test the validity of the RBC theory. These experiments have involved the recognition of partial objects, occluded objects, and incomplete objects by human subjects. Experiments have also been conducted to test the ability of people to recognize objects from simple line drawings as opposed to color photographs. The first series of experiments confirmed the importance of the non-accidental relationships and strengthened the hypothesis that people view objects as a connected group of components. The second set of tests confirmed that the recognition times for line-drawn objects are almost identical to those for color photographs of objects.

The RBC theory has added a very important aspect to models of human vision: that previously unknown objects can be subdivided into familiar components, and that their representations can reside in memory even before they are named. Such generic object representation capability has not been present in systems discussed to this point, which relied heavily on the application of detailed *a priori* knowledge for recognition. Biederman has suggested that the RBC theory could be used as the basis for a computational vision system, and someone has taken him up on that suggestion as will be seen in the following section.

8.5.3 PARVO vision system

Background

The PARVO computational vision model is based on Biederman's recognition by components (RBC) theory. PARVO stands for **primal access recognition of visual objects** (Bergevin and Levine, 1993). Primal access is defined as the conversion of the first contact with a perceptual input from an isolated, unanticipated object to a representation in memory (Biederman, 1985). An important feature of PARVO is this ability to classify generic unknown three-dimensional objects into general categories based solely on the structure of the object and the shape of its components. The information on the structure and shape of these three-dimensional objects is extracted from a two-dimensional line drawing representing a single view of the object. Unlike the systems previously described, PARVO does not rely on accurately specified internal models as a basis for recognition, but rather identifies objects as members of a class, where each class is represented by a coarse, qualitative model.

Object representation

PARVO utilizes a subset of the volumetric components suggested by Biederman in RBC. These components are generated in a similar manner to that used by Biederman, utilizing combinations of the non-accidental relationships as applied to generalized cone cross-sections and axes. The way in which the 11 basic components used by PARVO are generated is illustrated in Fig. 8.12. Once a component has been identified as belonging to one of these categories, the cross-sectional symmetry relationship is applied to further distinguish components. Because the symmetry relationship can take on one of two values in PARVO (symmetrical or asymmetrical), the total number of different components representable is 22. Most previous vision models have dealt either with only smooth objects (Barrow and Tenenbaum, 1981) or with only polygonal objects (Sugihara, 1986). By utilizing the RBC components, PARVO can handle a combination of both kinds of objects.

A given object is represented as a connected graph, in which the

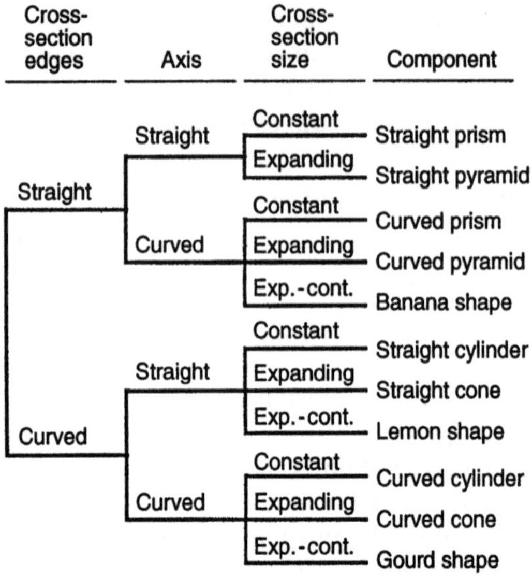

Figure 8.12 Volumetric components used in the PARVO vision system.

nodes of the graph represent the various components that make up the object, and the arcs represent the way in which those components are interconnected. Each node contains the component identification, as well as a symbolic variable representing the aspect ratio of the component. This variable can take on one of three values (stick, plate or blob) and represents the approximate appearance of the component (Shapiro *et al.*, 1984). The arcs of the object graph are labeled with the type of connection (end or side), as well as the approximate relative size of the two connected components. This technique allows the object representations to be properly proportioned while remaining scale invariant.

The models used by PARVO are represented in a similar fashion to the object graphs. Given a known object, an operator enters the correct component identifications manually into graph nodes, and the proper arcs are also created and labeled manually. One graph is created to represent each known model. This approach allows the model to represent the shape and structure of an object, while being imprecise on exact dimensions.

Object recognition

In the recognition process, PARVO starts with line drawings in which individual line and curve segments are already identified. The designers of PARVO sought to avoid the complications of edge detection and line and curve finding and segmentation, in order to concentrate on the higher recognition process. PARVO also works only with views of single objects, so the problems of segmenting different objects and the occlusion of one object by another are avoided. All of the components represented in the drawing are assumed to belong to a single object.

An object is first segmented into its constituent components. Separations are made at cusps and T-junctions, as recommended in RBC. Rules are then applied to each separated component to determine if the line and curve segments that make it up are edges of faces (the sides of GCs) or ends (the visible cross-sections of GCs). This rule-based technique is rather involved (Bergevin and Levine, 1992a), and it is not elaborated on further here. Once the face/end decision has been made, the component is classified into one of the 22 established categories based on the relationships previously described. The separations previously applied to divide the components are then analyzed in order to determine the connections between the components, and the other attributes (relative size, aspect ratio and type of connection) are computed. The complete object graph is then constructed accordingly. Figure 8.13 illustrates the component separation process, and the resulting graph, for the flashlight object.

The procedure used to match the object graph obtained from the image with the graphs representing the known models is complex (Bergevin and Levine, 1992b), and it will not be completely explained here. The procedure is essentially a two-level matching process. First, a coarse match is conducted which determines which models contain the combination of components present in the target object, and then a fine match is performed by determining the level of similarity between the

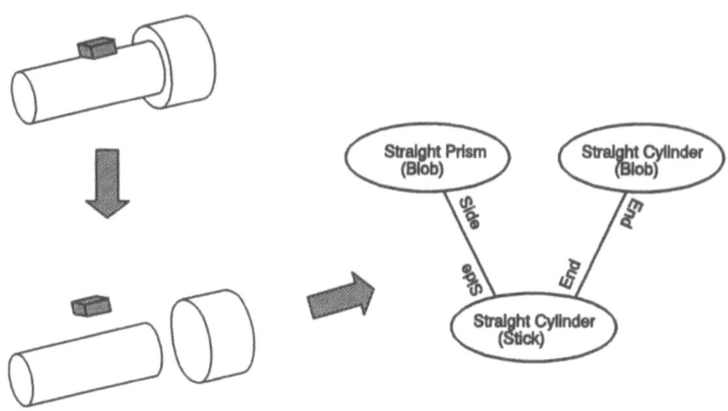

Figure 8.13 The flashlight object segmented and converted into an interconnected graph.

connections represented in the object and the model. The model that represents the best match is returned, as long as the quality of the match is above some minimum threshold.

PARVO implementation

The PARVO system was implemented in the LISP programming language on a computer specifically designed for symbolic processing. A database of 23 models was used, with the objects being similar to those used by Biederman to conduct his RBC experiments. The models represented rather simple man-made objects (suitcase, cup, lamp, table, etc.), and PARVO attempted to recognize them from simple line drawings. The line drawings were manually input as previously described.

Tests conducted with the line drawings resulted in 21 of 23 objects being identified properly. Additional tests were run in which objects were incompletely drawn, or corrupted by distorting components. Although these tests resulted in lower recognition accuracy, performance was relatively robust and indicated graceful degradation. The most common cause for failure was the misidentification of a component in the image, which had a rather disastrous effect on the model matching process.

PARVO summary

PARVO represents an attempt to implement Biederman's RBC theory in a computer. It is significant in that it is capable of generic object recognition in a bottom-up manner, without having to apply detailed *a priori* knowledge of models. It is also important in that it embodies a system of object representation that is general enough to describe a huge variety of objects, and yet is specific enough to allow relatively fine classification of those objects. Finally, PARVO continues the trend toward reliance on the laws of perceptual organization established by SCERPO, since it uses the non-accidental relationships to distinguish between component classes. Although PARVO was demonstrated on a relatively small database of simple man-made objects, it represents a significant step towards endowing computers with the ability to represent and identify generic objects.

8.6 SUMMARY

The computational vision models presented in this chapter represent the basis for the first generation of artificial vision systems. Taking advantage of contemporary developments in computer hardware, software and artificial intelligence, these systems were the first to attempt the implementation of image understanding on at least a limited scale. Although limited by the speed of available hardware, these systems laid the groundwork for future development in the field. The Marr model in particular served to define the field of artificial vision and introduced many concepts and terms that have become standards.

By introducing successively more processing based on human perceptual capabilities, the ACRONYM, SCERPO and PARVO systems generated performance that was increasingly human-like. Overlapping and then succeeding the development of these computational systems were systems that were even more biologically inspired, the connectionist vision models. Like artificial neural networks, connectionist vision systems simulate biological performance by duplicating the highly interconnected nature of the cells and structures found in biological vision systems. Three connectionist models for processing visual information are presented in the following chapter.

9 Connectionist approach to artificial vision

9.1 INTRODUCTION

The previous chapter dealt with models of vision that relied on algorithms, data structure and formulae to imitate human behavior. The models presented in this chapter take a totally different approach. The designers of these models seek to 'reverse engineer the competition' (Hillis, 1985), i.e. to imitate human capabilities by reproducing the biological structure of the human vision system and brain. In Chapter 4, artificial neural networks for pattern recognition and vision were discussed, finishing with the neocognitron and HAVNET networks which were specifically designed for visual pattern recognition. In this chapter, three biologically based connectionist models of the human vision system will be presented, each of which employs neural networks as component parts.

9.2 GROSSBERG VISION MODEL

9.2.1 Background

The first connectionist vision model to be discussed was developed by Grossberg and his colleagues (Grossberg, 1983; Grossberg, Mingolla and Todorovic, 1989), and its intent was the interpretation of two-dimensional images for the purpose of learning, and then recognizing, three-dimensional objects. Another goal of this model was to explain observed neurological data and human perceptual tendencies. To this end the model is very biologically plausible and it has led to many predictions about the structure and interconnection of neurons in the visual system, some of which have already been confirmed experimentally. A block diagram representing the overall structure of the model is shown in Fig. 9.1. The three main elements of the model are the boundary contour system (BCS), the feature contour system (FCS), and the object recognition system (ORS). Each of these is explained in detail in the following paragraphs.

9.2.2 Boundary contour system

The BCS is comprised of four levels of processing, each composed of many cells. In the first or lowest level, cells with oriented, oblong receptive fields that are sensitive to edges, shading gradients and texture changes receive signals directly from the retina. These cells are illustrated in Fig. 9.2. These cells exist in all orientations, and they are arranged in opposite-contrast pairs. The structure of the BCS is illustrated in Fig. 9.3, and the opposite-contrast cell pairs can be seen on the first level. Only three different orientations are shown in the figure for simplicity.

The oriented opposite-contrast pairs of the first level connect to oriented cells in the next level up. These cells are sensitive only to the orientation and amount of the gradient present in the receptive region,

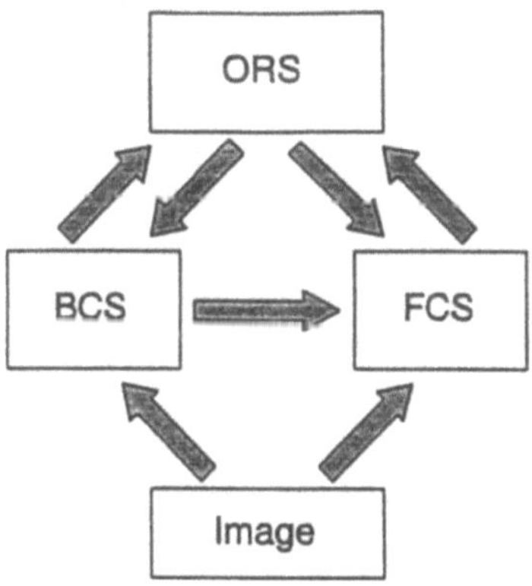

Figure 9.1 Block diagram of the Grossberg vision model.

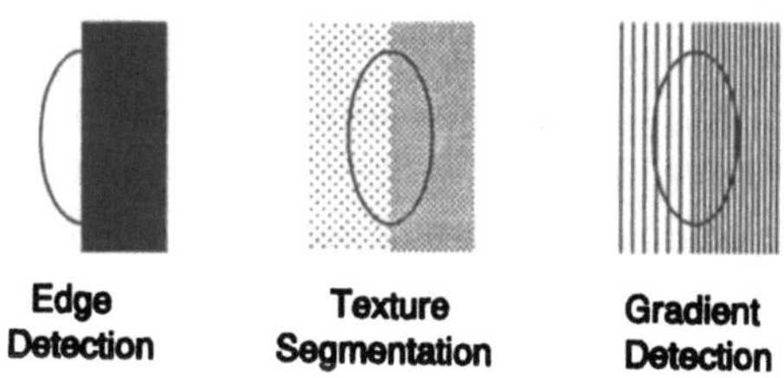

Figure 9.2 Oriented cells used in the boundary contour system.

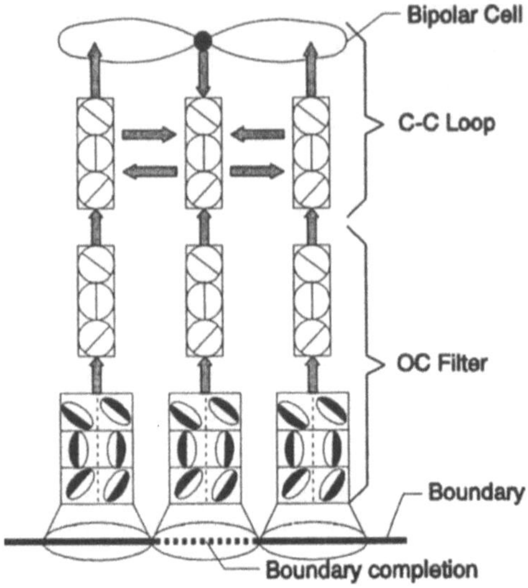

Figure 9.3 Structure of the boundary contour system.

not its direction. These first two levels of processing are called the oriented contrast (OC) filter, and they essentially perform a sophisticated form of edge detection. The outputs of the OC filter connect to cells on the third level which are also orientation sensitive. These cells engage in short-range competition, sending inhibitory signals to cells with the same receptive fields but with differing orientations (within the columns in Fig. 9.3), and to neighboring cells with similar orientations (between columns in the figure). The effect of these competitive interactions is to generate end cuts, which precisely define endpoints of edge segments.

The outputs of the third level connect to the fourth and final level of the BCS, which is comprised of cells with oriented receptive fields that extend over longer distances. These cells have a shape similar to a bow-tie, with long lobes extending from each side of center. These cells, called bipolar cells, exist at all orientations, but only one is shown in the figure for clarity. The bipolar cells perform a type of long-range cooperation, receiving signals in the lobes and transmitting signals from the center down to the previous layer. This cooperation has the effect of boundary completion, where a missing edge can be 'filled in' between like-oriented segments. This effect reproduces the Gestalt properties of continuation and collinearity. These last two levels of processing are called the competitive-cooperative (CC) loop.

Several computer simulations of the BCS have been conducted to demonstrate its properties (Grossberg and Mingolla, 1987). The BCS has been shown to perform well on edge detection, texture segmentation and smooth boundary completion. The output of the BCS is a boundary web, a net-like interconnection of edge segments of different strengths that separate areas of differing brightness or texture in the image.

9.2.3 Feature contour system

The boundary web is transmitted from the BCS to the FCS. The job of the FCS is to fill in the areas between the edges of the boundary web with signals representing the average brightness of the bounded region. The first layer of the FCS consists of cells that are sensitive to the brightness in the original image. The outputs of these cells are transmitted to a group of cells that comprise a **filling-in syncytium**, illustrated in Fig. 9.4. Signals in the filling-in syncytium rapidly diffuse across cell boundaries, except where inhibitory signals are received from the BCS. The diffusion is halted at these BCS boundaries, which coincide with the boundary web. The result of this processing is that the bounded regions defined by the boundary web are filled in to represent the average brightness of that area in the original image. Grossberg shows that areas of common brightness also represent areas of common depth, so this representation is analogous to the Marr 2½-D sketch (Chapter 8). Grossberg also states that the formation of this representation by the FCS is largely invariant of illumination conditions.

As with the BCS, the FCS has been extensively computer simulated. The BCS–FCS combination has been shown to create stable image representations, and several human perceptual phenomena like illusory contours and neon color spreading have been explained via BCS–FCS processes (Grossberg and Mingolla, 1985).

9.2.4 Object recognition system

The BCS and FCS processing just described occurs at several spatial frequencies. This is accomplished because the cells involved have

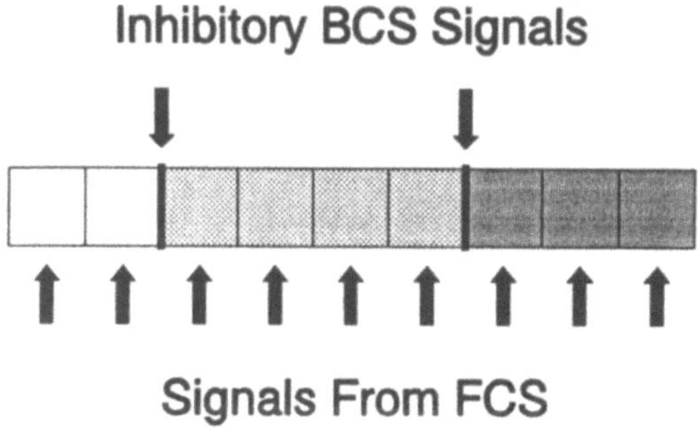

Figure 9.4 Behavior of the filling-in syncytium.

receptive regions of varying sizes. This allows the BCS–FCS to create a set of boundary and region features that are invariant to scale. The existence of cells at many orientations, as previously mentioned, allows for rotation invariance of these features. For object recognition, this invariant feature set is transmitted to an ART-like neural network (Carpenter and Grossberg, 1987a), the object recognition system (ORS). The ORS, like ART, self-organizes and recognizes objects from their features. The recognition process is interactive, with the ORS developing a hypothesis as to the identity of the object and transmitting this hypothesis back to the BCS and FCS. In the BCS and FCS this information is used to modify boundaries or regions based on previously learned information. The new BCS–FCS outputs are then sent to the ORS where the hypothesis is refined. This process is repeated until a stable object recognition is achieved.

9.2.5 Grossberg model summary

The BCS and FCS are theoretically very well developed, and several predictions based on their structure have been confirmed experimentally. Although the ORS is based on the well-developed adaptive resonance theory, the interactions between the ORS and the BCS–FCS are still somewhat nebulous. Also, the invariance of the model to translation, rotation and scale has not been thoroughly demonstrated. Finally, the Grossberg model has not been applied to problems involving cluttered scenes, partial or occluded objects and multiple objects.

Extensive computer simulations of certain aspects of the BCS and FCS have been conducted with impressive results, but very simple images were always used in these demonstrations. Furthermore, no complete, operational vision system has been constructed based on this model. As was the case with the neocognitron, the large number of cells and interconnections required for a complete model make practical implementation nearly impossible for a retina of any reasonable size. The Grossberg model has as its highest priority the imitation, explanation and prediction of biological phenomena, and its practical implementation into an artificial vision system was of secondary importance to its designers.

9.3 SEIBERT–WAXMAN VISION MODEL

9.3.1 Background

The vision model created by Seibert and Waxman (1992) resulted from a reversal of the Grossberg model's priorities. Practical applications were at the forefront of this model's design, and although biological

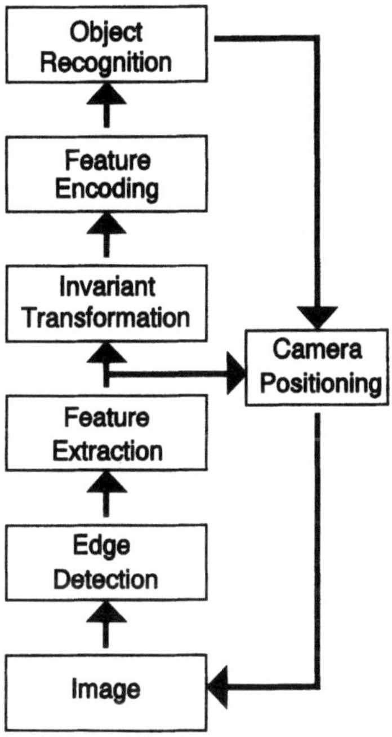

Figure 9.5 Processing stages of the Seibert–Waxman vision model.

plausibility was important to the designers, it was not an overriding consideration. Like the Grossberg model, the Seibert–Waxman (S–W) model contains several levels of processing. An interesting addition to this model is active camera position control, which corresponds to eye and head movement (visual saccades) in the biological sense. The structure of the S–W model is illustrated in Fig. 9.5. Each processing level is made up of multiple neuron-like processing elements, and some include a structure referred to as a **neural analog diffusion enhancement layer** (NADEL). The operation of each of the processing levels, and the inner workings of the NADEL, are explained in the following paragraphs.

9.3.2 Edge detection

The first processing level of the S–W model performs edge detection. The first step used in the edge detection process is binary thresholding, where pixels in the original image above a certain brightness threshold are replaced by white and those below the threshold are replaced by black. The edges are then detected using the difference-of-Gaussian (DoG) operator introduced in Chapter 3. The DoG operator is passed over the image, and the absolute value of the result for each pixel is calculated and passed to the next level of processing.

9.3.3 Feature extraction

In the feature extraction level the edges from the level below are smoothed and blended using a Gaussian filter. This filter is constructed in such a way as to cause areas of curvature in the edge to have high responses, with sharper curves and vertices returning the highest values. Local maxima are then found in this data, which correspond to locations of high edge curvature, vertices and line ends. These local maxima are the features that the model deals with in the object recognition process. The centroid of all of the features found is calculated in image (x–y) coordinates. This centroid information is fed to the camera positioning system, so the object can be centered in the image. The locations of the features and their centroid are also transmitted to the next processing level.

9.3.4 Invariant transformation

In this processing level, the x–y centroid from the previous level is used as the center of an object-based coordinate system. This step introduces translation invariance to the model. Using this centroid, the position of each feature is converted to log-polar coordinates. In the log-polar map, the angle of the feature from the centroid is plotted on the horizontal axis, and the log of the radius (distance from centroid to feature) is plotted on the vertical axis (Fig. 9.6). The result of this transformation is to convert scale and rotation variations to simple translations in the log-polar map (Schwartz, 1980). The centroid of the transformed features, in the log-polar map, is then calculated. Relative to this new centroid, the features are translation, rotation and scale invariant. The locations of the transformed features, and the log-polar centroid, are then transmitted to the next level.

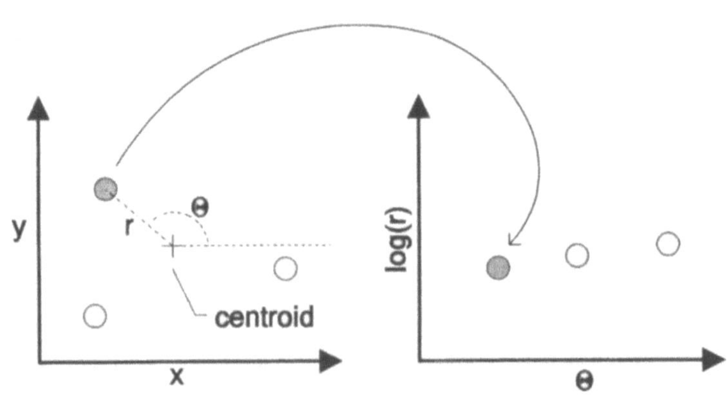

Figure 9.6 Illustration of the log-polar transformation.

9.3.5 Feature encoding

In this level a 5 × 5 grid of cells with overlapping receptive fields is projected over the features transmitted from the previous level, with the grid centered at the log-polar centroid. The cell at each grid point computes a value that is inversely proportional to the distance from its center to the nearest feature present in its receptive field. These values are then assembled into an analog valued feature vector with 25 elements, representing a single two-dimensional view (aspect) of the object in the image. Figure 9.7 illustrates this process. The feature vector is transmitted to the final processing level.

9.3.6 Aspect matching and evidential reasoning

At this level an ART-2 neural network is used to classify the feature vector into an object category. Full competition is not employed in the ART-2 output layer, so several nodes can respond at various levels. If two nodes are close to the maximum response, other evidence is used in order to determine the identity of the object.

Each object is represented by an aspect graph which represents the different learned views (aspects) of that object, and which indicates likely or possible transitions between aspects. For an object with N aspects, this graph is represented as an $N \times N$ matrix. Each element in the matrix represents the learned likelihood of the transition from the aspect represented by the row to the aspect represented by the column.

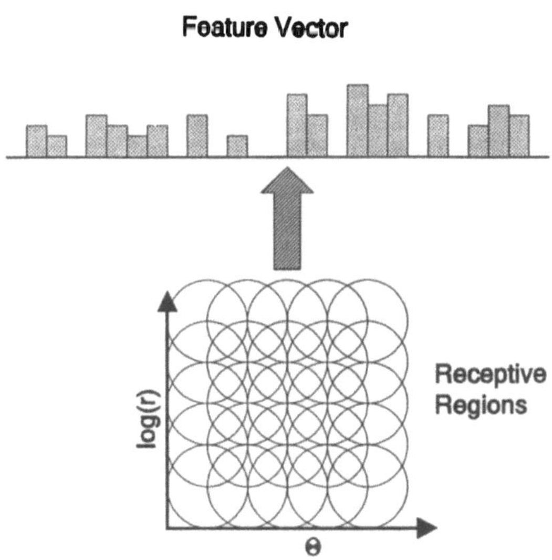

Figure 9.7 Feature encoding by overlapping receptive fields.

In the aforementioned case where two ART-2 nodes have nearly equal outputs, the aspect graphs for the objects in question are consulted. If one of the ambiguous nodes represents an aspect for the object last recognized by the model, and the transition between the two aspects is likely and consistent with known camera movement, enough evidence accumulates for this node and it wins out. If this is not true, and the other ambiguous node represents an aspect of a different object that is more consistent with camera position, evidence will accumulate and that node will win out. The exponential buildup and decay of evidence for each object results in a smooth transition between decisions, and a certain amount of hysteresis must be overcome in order to change the decision. If no unambiguous decision can be made, a camera movement will be commanded with the intent of eliminating the ambiguity.

9.3.7 Neural analog diffusion enhancement layer

Seibert and Waxman have developed the NADEL as a piece of multipurpose neural machinery that is biologically plausible, and is capable of performing many of the tasks necessary to implement their vision model (Seibert and Waxman, 1989). Figure 9.8 shows a block diagram for the NADEL. Note that cells internal to the NADEL exhibit the simple center–surround responses commonly found in the LGN and visual cortex (Chapter 3). Seibert and Waxman have shown that the NADEL is capable of edge smoothing, feature extraction, centroid location in the image and log-polar planes, and feature encoding.

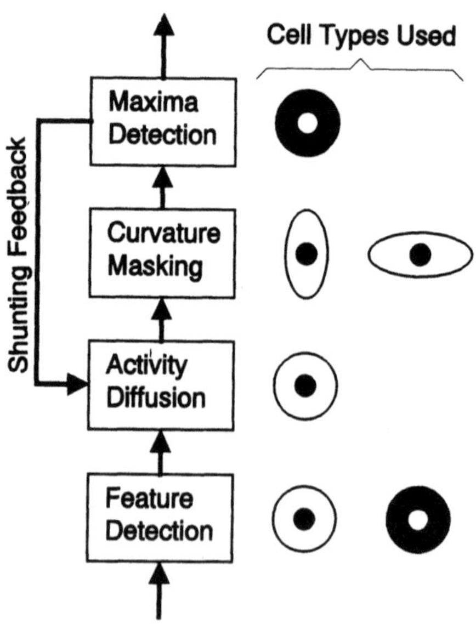

Figure 9.8 Processing stages of the NADEL.

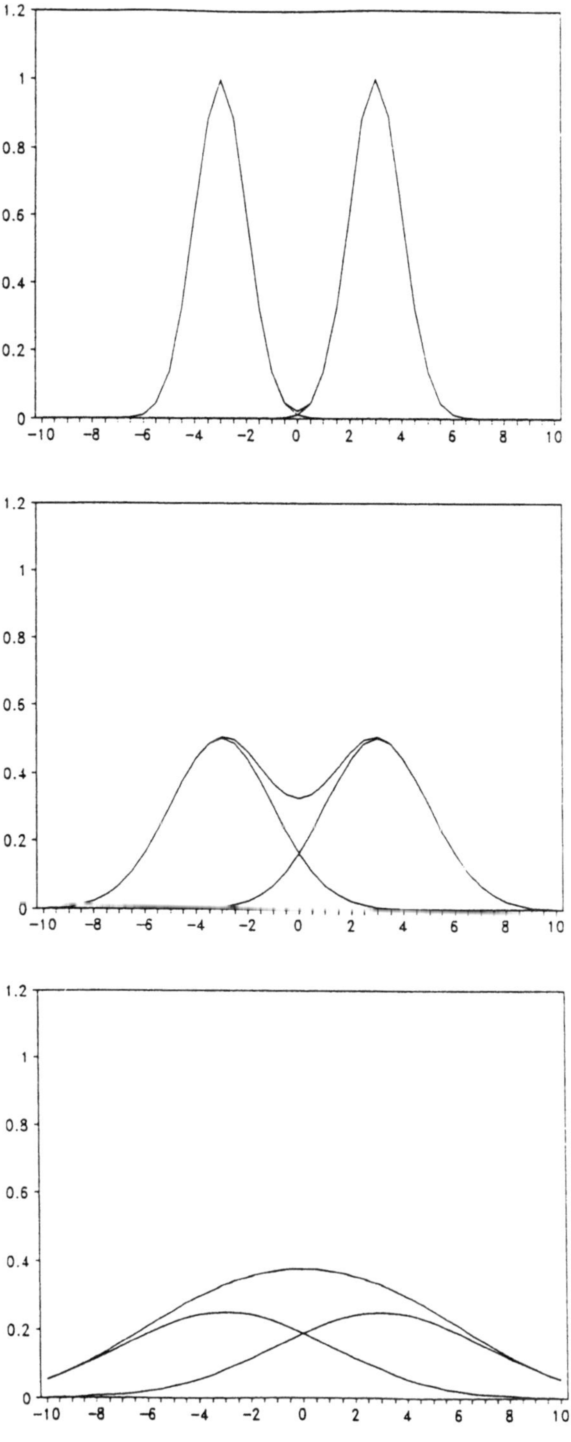

Figure 9.9 Illustration of NADEL diffusion over time.

The operation of the NADEL for centroid finding is presented next as an example. In the one-dimensional case, the NADEL initially creates a narrow Gaussian envelope of activity around each feature, and then the envelopes are allowed to spread and diffuse together. Figure 9.9 illustrates how the diffusion results in a single Gaussian envelope at the centroid after enough time has passed. In the two-dimensional case, the NADEL forms a narrow Gaussian spike of activity at each feature in the x–y plane, and then these are diffused into a wide Gaussian bubble of activity centered at the centroid of the features. Similar NADEL behavior proves advantageous in the edge smoothing, feature extraction and feature encoding tasks.

9.3.8 Seibert–Waxman model summary

Seibert and Waxman have developed a model of the vision system that is both practical and biologically plausible. The model has been simulated on a computer system, and applied to the task of recognizing airplanes from captured images (Seibert and Waxman, 1992). The images were taken with a video camera, and accurate scale model aircraft, painted flat black, were used in the tests. Although the aspect-graph object representation appears complex, Seibert and Waxman found that fewer than 20 aspects were required to represent each plane, and fewer than five transitions were likely from any given aspect. The simulation provided impressive recognition accuracy, even when a great deal of ambiguity was present in the views presented in the images.

Among the strengths of the S–W model are the fact that it has been completely implemented in a practical system, and that the system was capable of the intelligent recognition of three-dimensional objects from multiple two-dimensional views. Also, the S–W model is unique in that it includes active camera position control. One of the limitations of the system is the fact that the feature representation is not very rich, i.e. several objects could have very similar feature maps since only vertices and line ends are represented. Also, in testing the implementation, binary thresholded images of black painted objects were used. Analyzing natural images would prove far more difficult. Finally, the problem of cluttered scenes with multiple occluded objects was not addressed, as the simulation included only simple single-object scenes.

9.4 CAMERA VISION MODEL

9.4.1 Background

Building upon the research that has been presented thus far, and incorporating some novel ideas, a new connectionist model for artificial vision has been developed. The design and specification of this model are covered in the following sections. This new model has also been evaluated experimentally under various conditions, and those experiments are explained in detail in Chapter 10. Although this model has many similarities to the previously reviewed models, it also has several important differences and hopefully some improvements.

The motivation for the development of this model was simple: to create a practical artificial vision system that could recognize objects in a manufacturing environment. The goals addressed during the development of the model were as follows:

1. The model must be implementable as an artificial vision system using existing technology for a reasonable amount of money. If a vision system requires the processing power of a multimillion dollar supercomputer it is of little practical interest.
2. Reasonably dissimilar objects should be recognizable correctly under a wide variety of conditions. A vision system that recognizes only specific objects under highly restricted conditions is of little practical use.
3. Modestly complex objects should be recognizable in a reasonable amount of time. Individual objects should be identified in under 5 seconds if the system is to be of interest to manufacturing industries.

In addition to these primary design goals, it was desired that the model meet the following guidelines.

1. The system should be able to learn objects from examples rather than be based on pre-programmed object models. Introducing new objects to the system should be as simple as possible.
2. The model should employ processes that can be conducted in parallel as much as possible. Parallel processing is undoubtedly the future of computing for image processing, and the model should be able to take advantage of this emerging technology to further improve performance.
3. The processing steps employed in the model are to be as biologically plausible as possible. The existence of the mammalian vision system is too powerful to ignore.

The architecture of the vision model that was developed in order to attempt to meet these goals is shown in Fig. 9.10. The model is divided into three main stages. The first stage, representing early vision, consists of image acquisition and compression, edge detection, edge enhancement, vertex extraction and vertex connection. This stage extracts complex features from the input image, and passes them up to

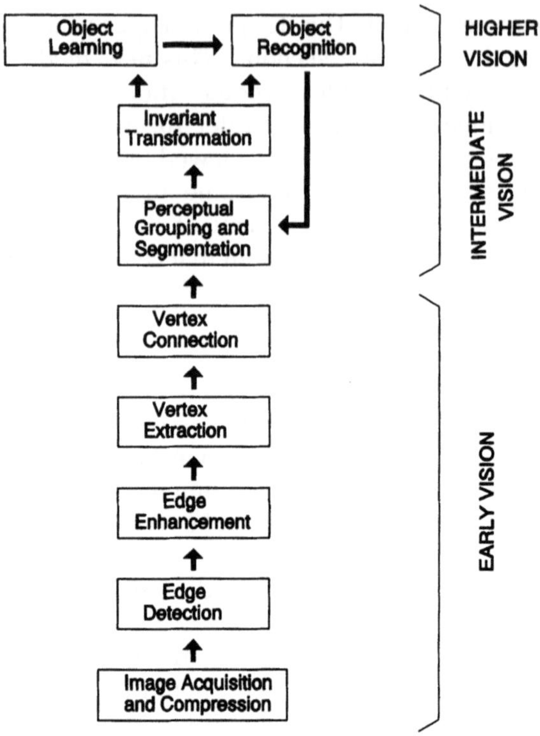

Figure 9.10 Processing stages of the CAMERA vision model.

higher levels of processing. The next stage, representing intermediate vision, groups and/or segments these complex features into objects, and performs invariant transforms on them in object-centered coordinates in order to prepare them for recognition or learning. The final stage, higher vision, performs learning and recognition on these transformed object representations. This stage employs a novel artificial neural network architecture, and object recognition is not based on pre-programmed models but rather on internal representations learned from experience.

Each stage of the model employs as much neural network-like parallel processing as possible, in order to improve performance and to maintain biological plausibility. This new model has been named the **connectionist architecture for multidimensional entity recognition using aspects**, or CAMERA. A detailed description of the design and operation of each stage of CAMERA is presented in the following paragraphs. The specific implementation described here was designed for use on a personal computer, but the specification for CAMERA is intended to be general enough to be implemented in many computing environments.

9.4.2 Early vision stage

Image acquisition

The first step in the early vision stage of the CAMERA model is the acquisition of the image that will be analyzed. In this implementation of CAMERA, images were acquired using a CCD video camera which was connected to an image acquisition board. This board was in turn installed in an IBM compatible personal computer (PC). The camera and card combination produced a raw image that was 640 × 480 pixels in 8-bit (0–255) grey-scale format. The specifications of the camera, card and computer are given in Chapter 10. Some images were acquired in ambient lighting conditions (indoors in a laboratory environment), and others were acquired using additional lighting.

Once the image was acquired, it resided in memory on the image acquisition (IA) card. The image was then read from the IA card, compressed by a 2:1 ratio in each dimension, and stored in the PC memory as an array of bytes. This compression was done simply because memory constraints in the PC would not allow storage and processing of the full raw image. The compression was accomplished by sampling the upper left corner of each 2 × 2 pixel area in the original image. This was done, rather than averaging, in order to preserve as much of the detail in the original image as possible. The result of this compression was a 320 × 240 image represented as an array of bytes, with values ranging from 0 to 255, in the PC memory. All further processing was then done on the compressed image. Figure 9.11 shows a typical compressed image.

Edge detection

Before going any further, it should be noted that in the mathematical description given next, and in all subsequent equations, a certain convention has been followed for referring to matrices or images and their elements. An upper case letter refers to a matrix or image as a unit (e.g. S), and the same letter in lower case (with subscripts) refers to an element of that matrix or image (e.g. s_{xy}).

The edge detection process employed by CAMERA proceeds by convolving the Sobel operator with the stored 320 × 240 image, and producing a new edge image which represents the edge strength values computed by the detector at each point in the original image. Although the Sobel operator is also known to provide a weak indication of edge orientation, only the edge strength information was utilized in this implementation. Because the operator consists of two 3 × 3 masks, it must be applied at least one pixel in from any image edge. For this reason the first and last rows and the first and last columns of the edge

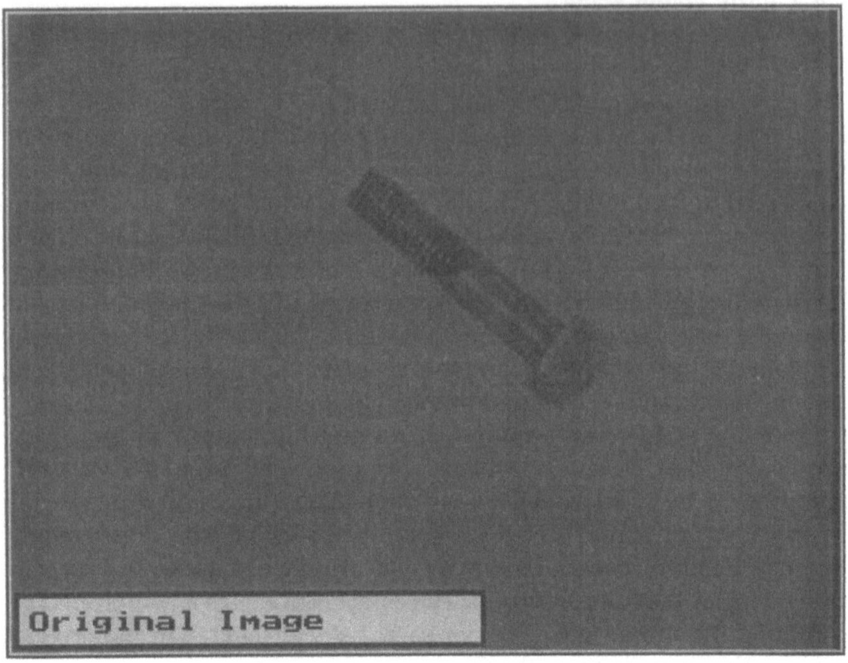

Figure 9.11 Example raw grey-scale image of a bolt.

are assigned values of zero as a first step. Next, the edge detector mathematically computes the values for interior points as follows:

$$s^0_{xy} = \beta \left| \sum_{i=0}^{2} \sum_{j=0}^{2} m_{(x+i-1)(y+j-1)} v_{ij} \right| + \beta \left| \sum_{i=0}^{2} \sum_{j=0}^{2} m_{(x+i-1)(y+j-1)} h_{ij} \right| \tag{9.1}$$

where M is the input image, β is the scaling factor, and S^0 is the Sobel edge image.

A scaling factor of $\beta = 1/6$ was used to restrict the output of the edge detector to the range 0–255, so a one-byte representation for each pixel could be used. The vertical (V) and horizontal (H) Sobel matrices are defined as follows:

$$V = \begin{bmatrix} -1 & 0 & 1 \\ -2 & 0 & 2 \\ -1 & 0 & 1 \end{bmatrix} \qquad H = \begin{bmatrix} -1 & -2 & -1 \\ -1 & 0 & 0 \\ 1 & 2 & 1 \end{bmatrix} \tag{9.2}$$

Once the edge values are computed, they are passed through a linear threshold to eliminate insignificant edges,

$$s_{xy} = \begin{cases} s_{xy}^0 & \text{if } s_{xy}^0 \geqslant \theta \\ 0 & \text{otherwise} \end{cases} \tag{9.3}$$

where S is the thresholded Sobel image.

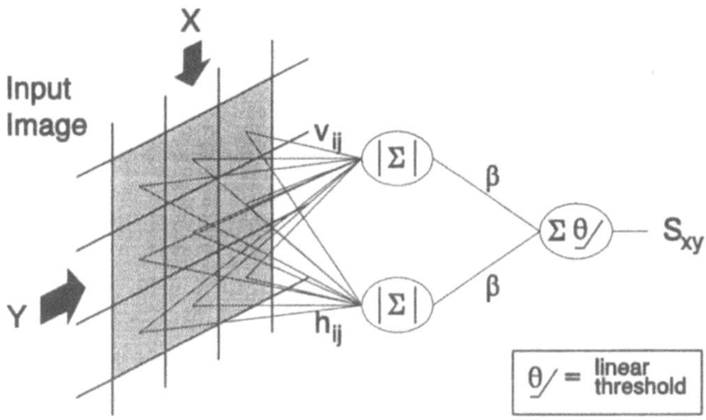

Figure 9.12 Architecture of the edge detection operator.

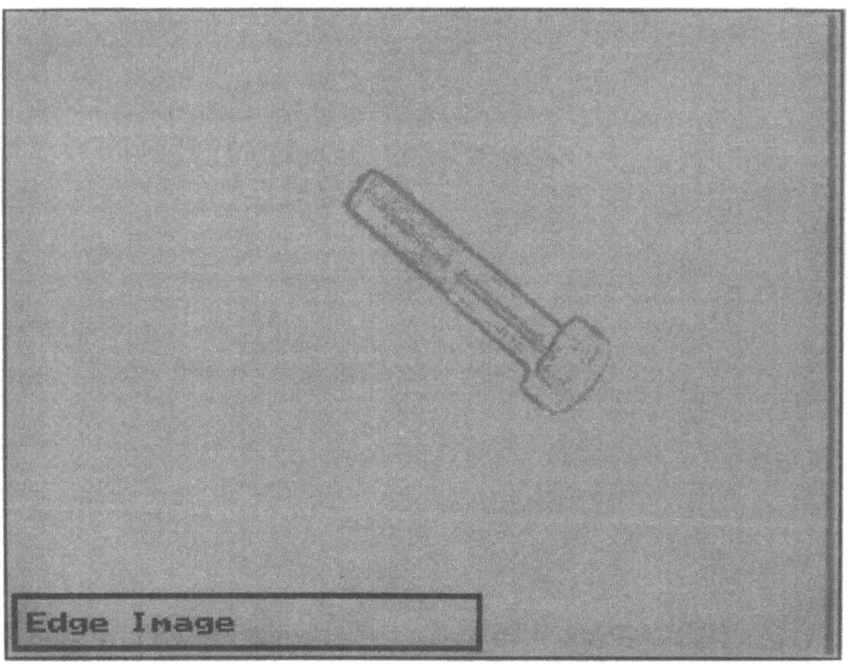

Figure 9.13 Edge image generated from the bolt image of Fig. 9.11.

A threshold value of $\theta = 16$ was used in this particular implementation of CAMERA. The architecture of the edge detection algorithm is shown in Fig. 9.12. Figure 9.13 shows the edge image that is the result of applying the edge detection algorithm to the image of Fig. 9.11.

Edge enhancement

As can be seen in Fig. 9.13, the output from the edge detection algorithm varies with the strength of the original edge. Weak edges generate narrow and rather light representations, whereas strong edges generate dark representations up to four pixels wide with a rather fuzzy appearance. It is desirable, prior to further processing, to convert these various edges into thinned binary representations. This is the job of the edge enhancement algorithm.

In order to illustrate how this enhancement is accomplished, it is necessary to introduce the concept of intensity plots. An intensity plot of an image is a three-dimensional graph in which the x and y dimensions represent the x and y coordinates of pixels in the image, and in which the z dimension represents the intensity of the pixels. Consider the image in Fig. 9.14. The square region outlined in the image represents a

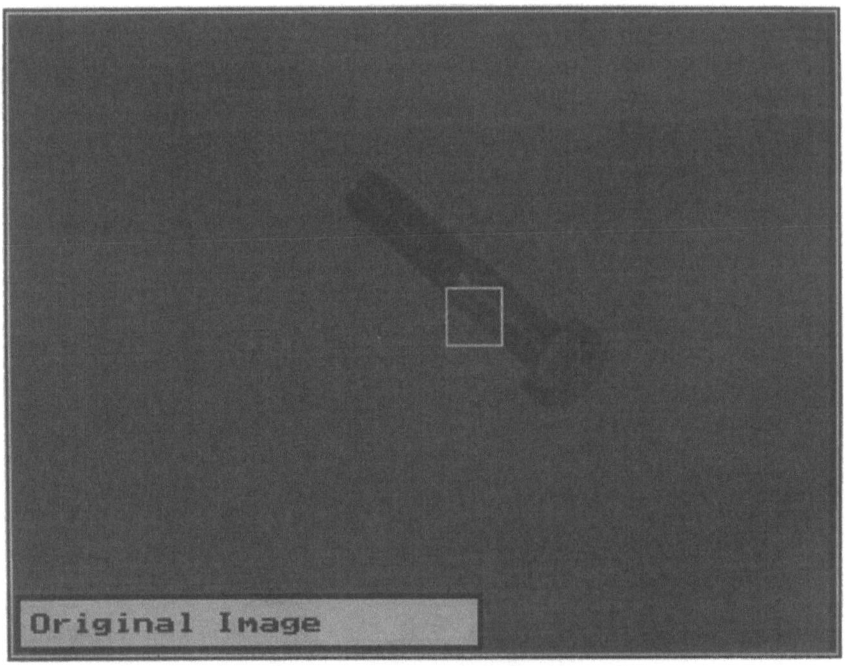

Figure 9.14 Region selected for intensity plots.

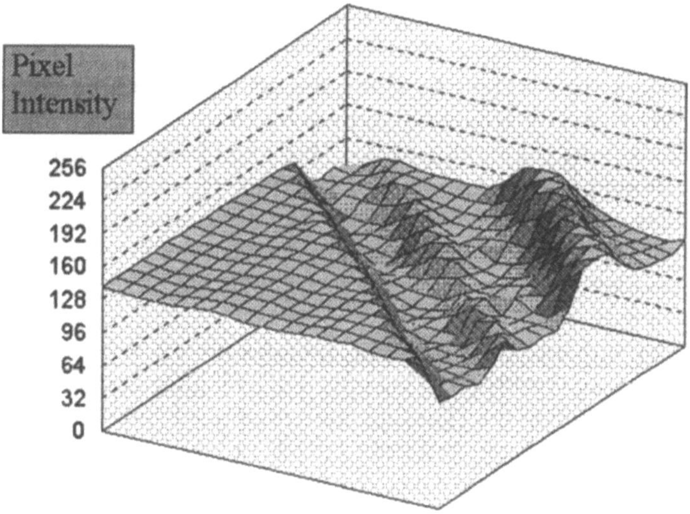

Figure 9.15 Image intensity plot of the area selected in Fig. 9.14.

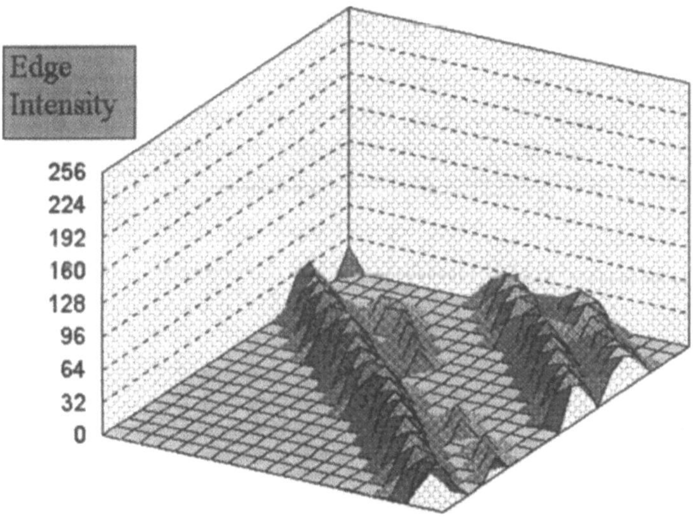

Figure 9.16 Edge intensity plot of the selected area.

20 × 20 pixel area. Figure 9.15 shows an intensity plot for that particular area of the image. Note that in the plot, edges are indicated by step-like structures.

Figure 9.16 shows the intensity plot for the corresponding area in the edge image generated from the image in Fig. 9.14, with the vertical axis representing the output of the edge detector. The step edges in the original image have become ridges in the edge image, with the height of the ridge corresponding to the relative strength of the associated edge. The peaks of these ridges lie along the actual location of the original edges. In order to accurately represent these detected edges it is desirable to locate these ridge peaks. This could be accomplished simply by thresholding the edge image, but this technique would delete some weak edges and it would still represent strong edges as being wider than one pixel. For this reason an edge enhancement operator was developed that would enhance ridges in the edge diagram while suppressing all other pixels.

The edge enhancement operator operates on a 3 × 3 neighborhood surrounding each pixel, and it is applied in a manner similar to the Sobel edge detection operator. The operator is applied only to the interior pixels of the edge image, and not to pixels on the extreme outer edges. This is of no consequence since those pixels were deliberately zeroed during the edge detection process. Mathematically, the edge enhancement operator is applied to the edge image as shown below:

$$e^0_{xy} = s_{xy} + \gamma \sum_{i=0}^{2} \sum_{j=0}^{2} \operatorname{sgn}(s_{xy}, s_{(x+i-1)(y+j-1)}) u_{ij} \tag{9.4}$$

where γ is the sensitivity multiplier and E^0 is the enhanced edge image. A multiplier of $\gamma = 20$ was used to control the sensitivity of the enhancement process. The function sgn(a,b) computes the algebraic sign of the difference between a and b as follows:

$$\operatorname{sgn}(a,b) = \begin{cases} 1 \text{ if } a > b \\ 0 \text{ if } a = b \\ -1 \text{ if } a < b \end{cases} \tag{9.5}$$

The edge enhancement matrix U is defined as

$$U = \begin{bmatrix} 2 & 1 & 2 \\ 1 & 0 & 1 \\ 2 & 1 & 2 \end{bmatrix} \tag{9.6}$$

Once the enhanced edge values are computed, they are converted to binary values using the threshold function

$$e_{xy} = \begin{cases} 1 \text{ if } e^0_{xy} > \phi \\ 0 \text{ otherwise} \end{cases} \tag{9.7}$$

where E is the thresholded-enhanced edge image. The threshold value of $\phi = 127$ (50%) was used in this particular implementation.

The edge enhancement operator is differential in nature, that is it operates on the algebraic sign of the difference in value between a pixel and its neighbors. In this way pixels that are higher than most of their neighbors are enhanced, while those that are not are suppressed. This enhances ridges and peaks while suppressing pixels on slopes or flat areas. The effect of the enhancement can be seen in Fig. 9.17, which shows an intensity plot of an area of an enhanced edge diagram that corresponds with the area plotted in Figs. 9.15 and 9.16. The architecture of the edge enhancement algorithm is shown in Fig. 9.18. Figure 9.19 shows the enhanced edge image produced from the original edge image of Fig. 9.13.

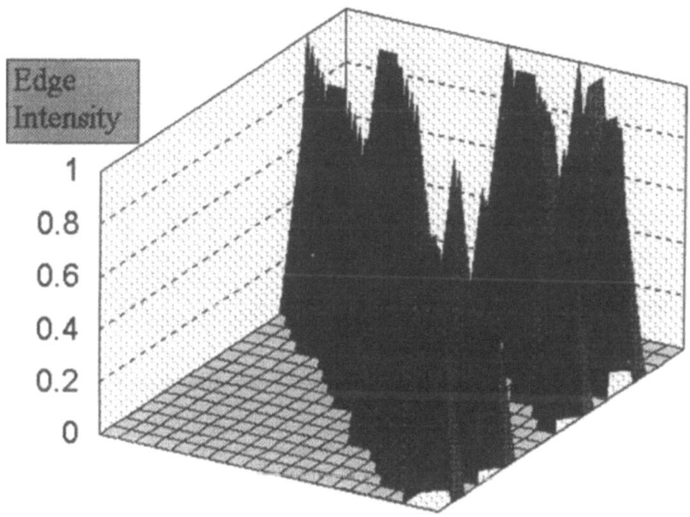

Figure 9.17 Enhanced edge intensity plot of the selected area.

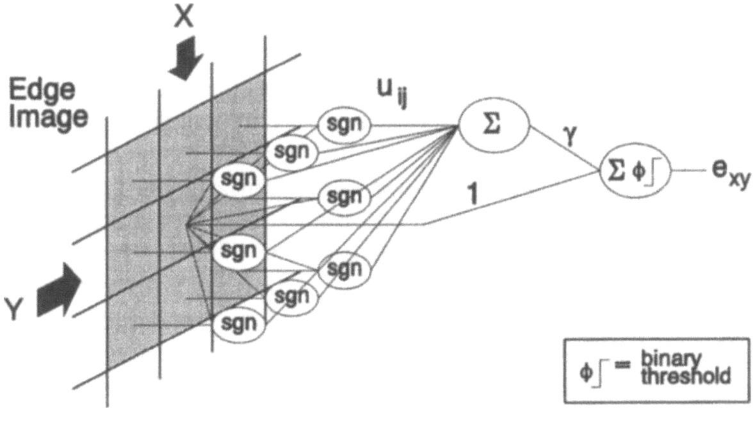

Figure 9.18 Architecture of the edge enhancement operator.

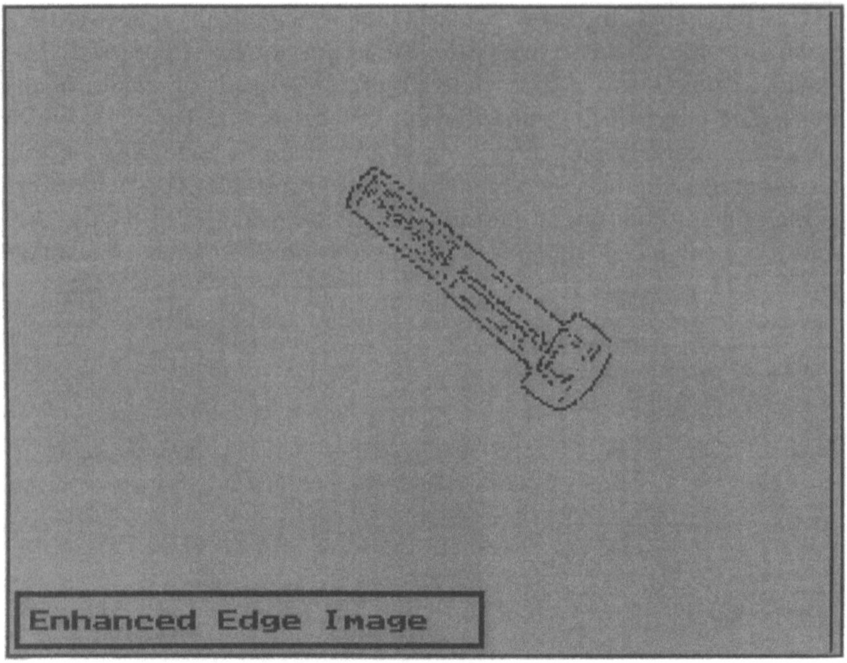

Figure 9.19 Enhanced edge image generated from the edge image of Fig. 9.13.

Vertex extraction

Although the enhanced edge image is precise and rich with information about the original image, edge points themselves are not features that lend themselves to efficient manipulation. Edge images are not easily scaled, rotated or transformed, and if an image contains multiple objects the edge representation is difficult to divide into multiple entities. These are some of the reasons that the next processing step implemented in CAMERA performs vertex extraction. In this context a vertex is defined as either a point of high curvature or an intersection point in the enhanced edge image.

There exists much biological evidence relating to the importance of vertices as visual features. From children's dot-to-dot drawings to the interpretation of celestial constellations as objects, humans have demonstrated the ability to convert vertex representations into meaningful line drawings of objects. Attneave (1954) demonstrated the usefulness of curvature extrema along contours for two-dimensional shape representation and recognition. Biederman (1985) showed that, when recognizing partial line drawings, humans have good success when the missing sections are segments between vertices, but that they encounter difficulty when one or more vertex regions are deleted. This would suggest a heavy reliance on vertices as features in the human visual system.

Recently developed artificial vision models have also relied on vertices as important features. The recognition by components theory (Biederman, 1985), and the associated PARVO vision model (Bergevin and Levine, 1993), used T-junction vertices to segment objects into components. The Seibert–Waxman model relied on vertices as the only features upon which object recognition was based. Finally, recent enhancements to the Grossberg vision model suggested by Lehar (1993a;b) introduce vertices as primary representational features.

During vertex extraction in CAMERA, the enhanced edge image is first divided into 4 × 4 pixel areas. For each of these areas a 3 × 3 vertex extraction operator consisting of two kernals is passed over all of the pixels in the region. The output of this operator is only computed if the center pixel under the kernal is marked as an edge in the enhanced edge diagram. Of all the enhanced edge points found in the region, the one generating the greatest output is extracted as a vertex if that output exceeds a threshold value. If no edge points generate outputs that exceed the threshold, the last edge point visited in the region is extracted as a vertex anyway. This behavior guarantees that even perfectly straight edges produce vertices spaced at approximately four-pixel intervals. This fact is important to later processing, as will be seen in the following subsection. Only if a region has no edge points will it fail to produce a vertex, and no more than one vertex is produced by any region.

The two kernals used in the vertex extraction operator are in fact identical to the Sobel kernals. The differential properties of the Sobel masks detect an 'imbalance' under the operator. Because the operator is evaluated only when centered over an edge pixel, only a perfectly straight edge will be balanced in all directions. The operator is very sensitive to points of curvature, corners and intersection points. Figure 9.20 shows four typical edge patterns and the associated outputs generated by the vertex extraction operator. Because of the manner in which lines are represented digitally, many false vertices can be detected by the operator. The extraction process limits each region to a single vertex to prevent the proliferation of these false indications.

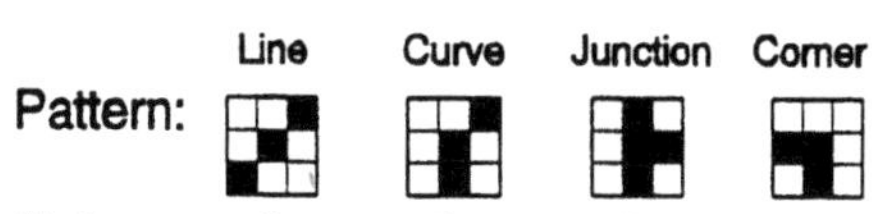

Figure 9.20 Vertex extraction operator outputs for various patterns.

Mathematically, the output of the vertex extraction operator is computed as follows:

$$r_{xy} = e^{T}_{xy}\left|\sum_{i=0}^{2}\sum_{j=0}^{2} e^{T}_{(x+i-1)(y+j-1)}v_{ij}\right| + e^{T}_{xy}\left|\sum_{i=0}^{2}\sum_{j=0}^{2} e^{T}_{(x+i-1)(y+j-1)}h_{ij}\right| \tag{9.8}$$

where E^T is the thresholded enhanced edge image and r_{xy} is the output for the vertex operator at point (x,y).

The vertical (V) and horizontal (H) matrices are the Sobel matrices previously defined (equation 9.2). The architecture of the vertex extraction operator is shown in Fig. 9.21.

Once the vertex extraction operator values are computed for an area A, the maximum is found as

$$(r_{xy})^* = \max\{r_{xy}\,|\,(x,y)\in A\} \tag{9.9}$$

A vertex (x,y) is then identified for extraction as follows:

$$(x,y) = \begin{cases} (a,b)\text{ if } (r_{ab})^* > \Gamma \\ (c,d)\text{ if } (r_{ab})^* \leqslant \Gamma \\ \quad\text{and } \{e_{cd}|e_{cd}\in A \text{ and } e_{cd}=1 \text{ and } c=c^* \text{ and } d=d^*\} \neq \emptyset \end{cases} \tag{9.10}$$

where Γ is the minimum vertex value threshold and (c^*,d^*) are the coordinates of the last enhanced edge point visited in A. A threshold value of $\Gamma = 1.5$ was used in this particular implementation of CAMERA. If neither of the conditions in equation 9.10 are satisfied, then no vertex is extracted from A.

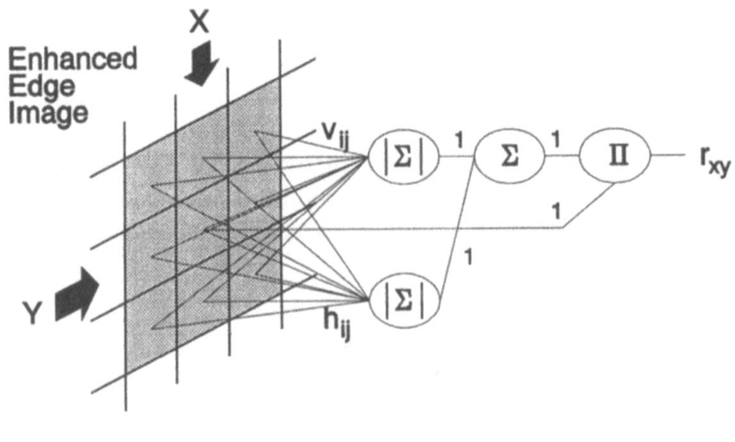

Figure 9.21 Architecture of the vertex extraction operator.

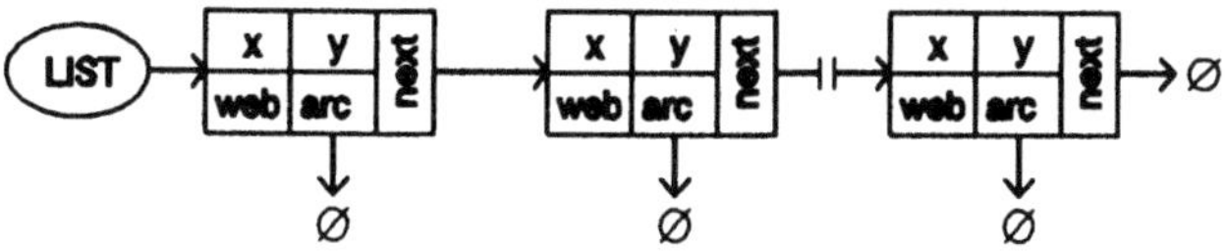

Figure 9.22 Vertex-list data structure.

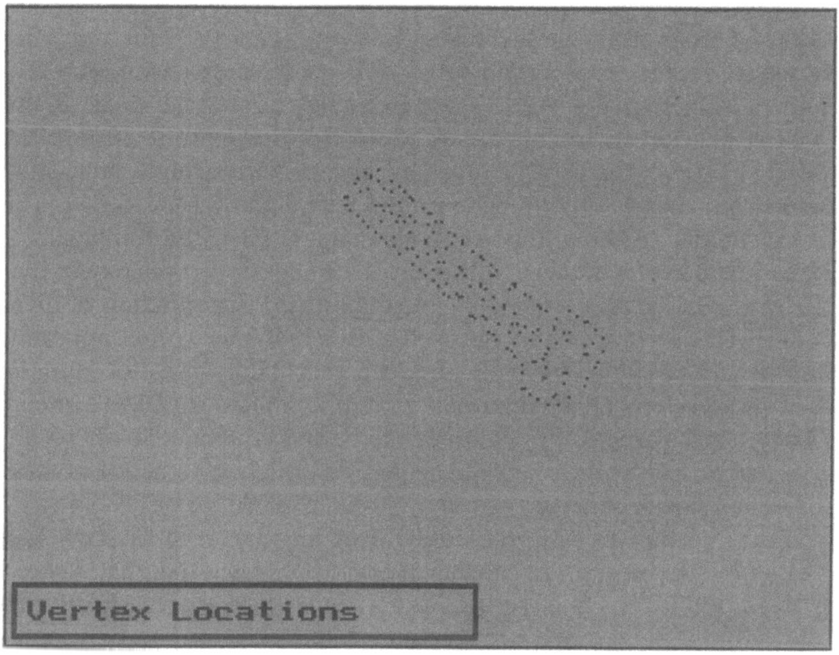

Figure 9.23 Vertices extracted from the enhanced edge image of Fig. 9.19.

Once a vertex is extracted it is inserted into a linked-list data structure which contains all of the vertices extracted for a given image. The vertex list data structure is illustrated in Fig. 9.22. Note that the web numbers and arc pointers, which will be utilized in later processing, are initially set to zero and null values, respectively. When this list has been built it can be traversed and the vertex locations can be displayed. Figure 9.23 is an image generated in this manner that shows the vertices that were extracted by CAMERA from the enhanced edge image of Fig. 9.19.

Vertex connection

Although the vertices themselves represent much of the information in the original image, the representation becomes much richer if the appropriate pairs of vertices are connected by short line segments. In order to determine if two vertices should be connected, the area between them in the enhanced edge image must be evaluated in some manner. If evidence is found for the existence of an edge between these vertices, some representation of a line segment between them should be established. This is the task which is accomplished by the vertex connection algorithm in CAMERA.

Because there is a possibility of a large number of vertices being extracted from an image (even simple images generate hundreds), the vertex connection process can suffer from combinatorial explosion. The main reason that the vertices are extracted from small areas in the manner previously described is to avoid this problem. Because of this local behavior of the vertex extraction process, vertex connection can be carried out over a small neighborhood surrounding each vertex. This fact gives rise to three important advantages. First, the combinatorial explosion is avoided, since only a small number of vertices (fewer than 10) can exist in the immediate neighborhood surrounding a given vertex. The second advantage is the fact that the vertex-connecting process can be done efficiently and in parallel, since the procedure for each vertex relies only on information from its immediate neighborhood. Finally, since vertices are guaranteed to be spaced rather closely on any given edge, curves can be represented accurately as a series of short connected line segments.

The connection process proceeds by searching the vertex list previously created for all pairs of vertices that are less than some specified distance apart. Specifically, two vertices $v_{x1,y1}$ and $v_{x2,y2}$ are candidates for connection if the following inequalities both hold:

$$|x_2 - x_1| < \lambda \qquad (9.11)$$

$$|y_2 - y_1| < \lambda \qquad (9.12)$$

A threshold value of $\lambda = 10$ was used in this particular implementation of CAMERA. Vertex pairs that meet the above conditions are tested for connection by generating an imaginary line between them and classifying the slope of that line into one of four categories: vertical, horizontal, 45-degree up and 45-degree down. Based on this classification, a 3×3

2	4	2
3	4	3
2	4	2

Vertical

2	3	2
4	4	4
2	3	2

Horizontal

4	3	1
3	4	3
1	3	4

45-degree down

1	3	4
3	4	3
4	3	1

45-degree up

Figure 9.24 Vertex connection templates for various orientations.

kernal is then applied to the enhanced edge image at a point midway between the two vertices. The kernals used for each classification are illustrated in Fig. 9.24.

If the value generated by the chosen kernal is above a threshold, enough evidence exists for an edge between the two vertices, and a connection is generated between them. Mathematically, the slope m of a line between two vertices $V_{x1,y1}$ and $V_{x2,y2}$ is computed as follows:

$$m = \frac{y_2 - y_1}{x_2 - x_1} \tag{9.13}$$

The vertex connection matrix C is then selected as

$$C = \begin{cases} C_v & \text{if} |m| > \tan(67.5°) \\ C_h & \text{if} |m| < \tan(22.5°) \\ C_d & \text{if } \tan(22.5°) \leqslant |m| \leqslant \tan(67.5°) \text{ and } m > 0 \\ C_u & \text{otherwise} \end{cases} \tag{9.14}$$

where: C_v if$|m|>\tan(67.5°)$
C_h if$|m|<\tan(22.5°)$
C_d if $\tan(22.5°) \leqslant |m| \leqslant \tan(67.5°)$ and $m>0$
C_u otherwise

A connection is established between the two vertices if the following inequality is satisfied:

$$\sum_{i=0}^{2} \sum_{j=0}^{2} e_{(x_c+i-1)(y_c+j-1)} c_{ij} > \mu \tag{9.15}$$

where the point (x_c, y_c) is a point midway between the two vertices and is found as

$$x_c = \left| \frac{x_1 + x_2}{2} \right| \qquad y_c = \left| \frac{y_1 + y_2}{2} \right| \tag{9.16}$$

A threshold value of $\mu = 300$ was used in this particular implementation of CAMERA.

Once a vertex pair is designated for connection, the pair is connected in the vertex list data structure by generating a bidirectional pair of arcs between the vertices. If the vertex pair (a,b) is to be connected, an arc $a \rightarrow b$ and an arc $b \rightarrow a$ are generated. An arc essentially consists of a pointer to the connected vertex, and a linked list of arcs is attached to each vertex. The arc list for a particular vertex contains arcs to each of the other vertices to which it is directly connected. Figure 9.25 shows the vertex list data structure expanded to include the arc lists.

Once all of the vertex connections have been established, the expanded vertex list is traversed to determine which groups of vertices

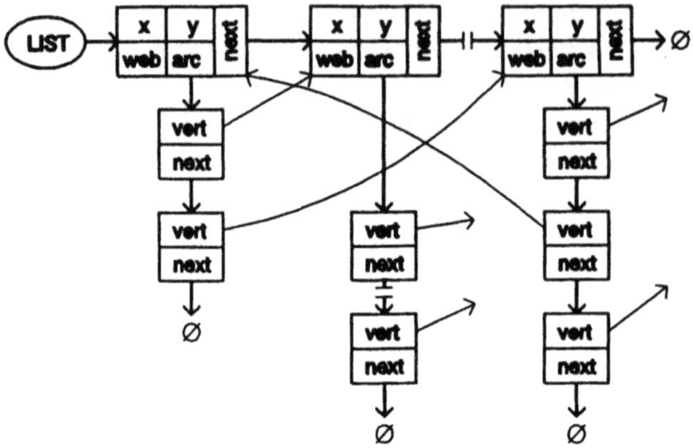

Figure 9.25 Vertex list data structure expanded to include arc lists.

belong to interconnected webs. A **web** is defined as the set of all vertices that can be reached by beginning at one vertex and traveling along all possible combinations of interconnecting arcs. The web is so named because it corresponds roughly to the boundary web produced by the boundary contour system (BCS) of the Grossberg model. The web building process begins by setting the current web number to 1, and starting at the first vertex in the list and marking this vertex as a member of the web by setting the web field in the vertex data structure to the current web number. All of the vertices directly connected to that vertex are then visited by following arcs, and their web numbers are set to the current web number. All of the vertices directly connected to each of these are then visited and their web numbers are also set to the current web number, and this process continues recursively until no more connected vertices are found that have zero web numbers. When this occurs, the current web number is incremented, the list is searched for any vertex with a zero web number, and the web building process begins again at that vertex. Web building is complete when all of the vertices in the vertex list have non-zero web numbers. Figure 9.26 illustrates the web building process for a simple set of interconnected vertices.

When web building is concluded, the result is that the vertex list is grouped into complex features that represent interconnected webs of edge segments. Note also that, because of the manner in which the webs were built, each vertex belongs to one and only one web. It is possible for the entire image to be represented by a single web, but it is far more likely that multiple webs will be generated. In an image with multiple separated objects, each web usually represents an object or a component of an object. In an image with multiple overlapping objects, a single web

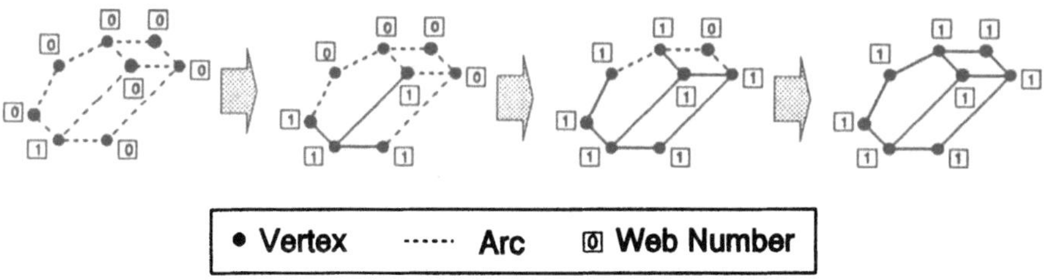

Figure 9.26 Illustration of how web building progresses.

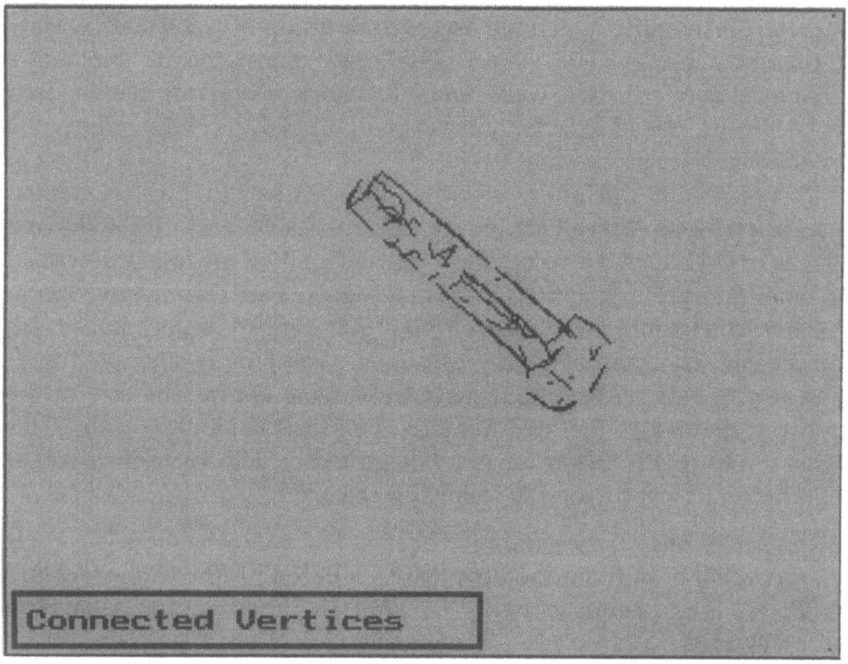

Figure 9.27 Web diagram for the bolt image of Fig. 9.11.

may represent more than one object. The webs are the complex features that are passed on to the intermediate vision stage of CAMERA. Figure 9.27 shows the webs extracted from the enhanced edge image in Fig. 9.19, by connecting the vertices shown in Fig. 9.23.

9.4.3 Intermediate vision stage

Perceptual grouping and segmentation

The first step in the intermediate vision stage is perceptual grouping and segmentation. Because it is possible for the webs generated by the previous stage to represent components of objects or multiple objects, some mechanism must be provided which joins webs that represent components of a single object in some manner in order to make recognition possible. Also, some mechanism must be provided which divides webs that represent multiple objects at appropriate separation points into two or more webs. The techniques that have been developed to accomplish these two tasks in CAMERA are intended to model, at least partially, the Gestalt laws of visual organization in human beings (Chapter 3).

To facilitate processing at this point, CAMERA uses the entity data structure. An entity is defined as a web or group of webs that possibly represent a recognizable object. The entity data structure contains a linked list of one or more webs, and it has storage locations that are used to keep track of a scale factor and a center point for the entity, which will be used in later processing.

To begin with, one entity is formed for each web that was generated by the previous stage. Each entity is then passed on to the following stages of CAMERA for recognition (the recognition process is described in detail later). Those entities that are recognized are then removed from further consideration. Those entities that remain unrecognized are candidates for grouping or segmentation. Grouping is attempted first, and it proceeds by beginning with the largest entity. The size of the entity is defined as the total number of vertices in all of its webs. This large entity (E_a) is tested for possible grouping with each of the other entities (E_b) by applying the following rules:

1. *Proximity rule* Group (E_a,E_b) if at least one pair of vertices (V_a,V_b) is separated by a distance of less than ε, where V_a is the vertex in entity E_a, V_b is the vertex in entity E_b, and ε is the maximum separation parameter.
2. *Containment rule* Group (E_a,E_b) if the center of E_b is within a distance r_{max} of the center of E_a, where r_{max} is the distance from the most outlying vertex of E_a to the center of E_a.

Each of these rules is in turn applied to all possible combinations of entities. Although this process can suffer from combinatorial explosion, the number of entity pairs that satisfy these rules is typically small, so the problem is minimized. When an entity pair is identified for grouping, this is accomplished simply by linking together the web lists represented in the paired entities. After each successful grouping, the resulting entity is passed up the processing hierarchy for recognition, and any time an entity is successfully recognized it is removed from further consideration. Grouping via the proximity rule is illustrated in Fig. 9.28, and grouping via the containment rule is illustrated in Fig. 9.29.

Once all possible grouping combinations have been tried, any

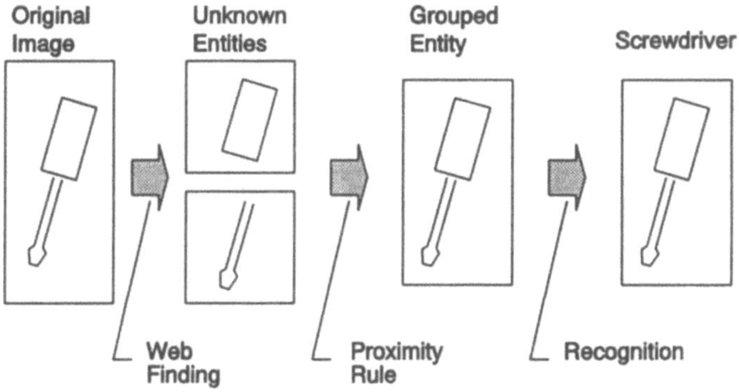

Figure 9.28 Entity grouping by the proximity rule.

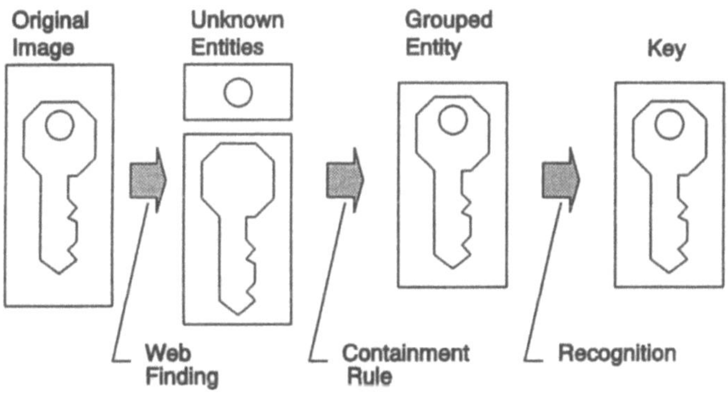

Figure 9.29 Entity grouping by the containment rule.

remaining entities are candidates for segmentation. An entity is segmented by identifying vertices that represent T-junctions (T-junctions are defined as in section 8.5.2), deleting these vertices and any associated non-collinear connections from the affected web, and testing the web to see if it has been divided into two webs. If it has, the segmented portion is used to create a new web, which is then used to create a new entity. Both of the segmented entities are then passed up the hierarchy for recognition. The process of segmentation is illustrated graphically in Fig. 9.30, where a single entity representing overlapping objects is divided into two entities, each representing an object. As before, any entities that are successfully recognized are removed from further consideration. When all possible segmentations have been performed, any remaining entities are simply identified as unknown.

Although the grouping and segmentation process may appear to be simply an image processing and data manipulation task, there is in fact a biological basis for this type of system behavior. Recent research

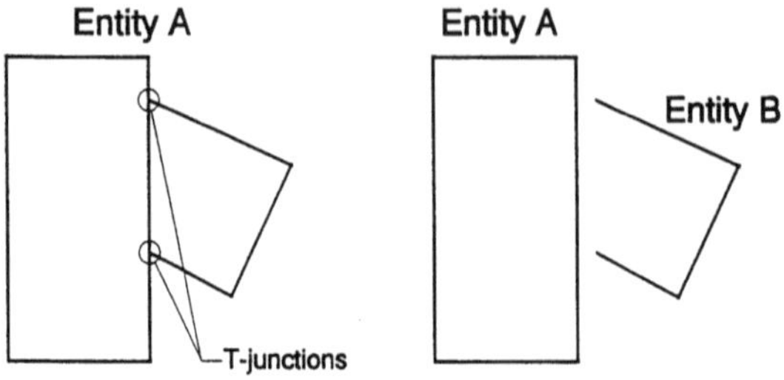

Figure 9.30 Example of entity segmentation at T-junctions.

developments in this area suggest that features such as the vertices used by CAMERA can be recognized by biological neurons, and that when these neurons fire synchronously with each other the features represented are grouped by the brain for recognition (Konig and Schillen, 1991). The synchronous behaviour has been observed to change over time, indicating that various groupings are tried before recognition is successful. It is this behavior that CAMERA is attempting to model in the perceptual grouping and segmentation phase.

Invariant transformation

An entity representation must be transformed in order to prepare it for recognition. The goal of this transformation is to provide shift, scale and rotation invariance to the recognition process. The first step of this process converts the vertices into polar coordinates, but before this conversion can take place the center of the entity must be found. The most desirable center point for the purpose of rotation invariant transformation is the center of the minimum spanning circle of the vertex set. The minimum spanning circle is defined as the smallest circle such that all of the vertices that are part of an entity lie either on the circle or in its interior (Melville, 1985). CAMERA employs a unique iterative approach to locate the center of the minimum spanning circle. First, the center point (x_c, y_c) of the entity is estimated as

$$x_c = \frac{x_{\min} + x_{\max}}{2} \tag{9.17}$$

$$y_c = \frac{y_{\min} + y_{\max}}{2} \tag{9.18}$$

where: $x_{\min}$ = minimum x-coordinate of all vertices in entity
$x_{\max}$ = maximum x-coordinate of all vertices in entity
$y_{\min}$ = minimum y-coordinate of all vertices in entity
$y_{\max}$ = maximum y-coordinate of all vertices in entity

Then the following procedure is employed to refine the estimate of the location of the center point:

1. Compute the maximum distance r_{max} between the estimated center point and the vertex in the entity most distant from it.
2. Identify the two vertices furthest from the present estimated center point, and locate a point midway between these two vertices.
3. Move the center point a short distance toward this midway point.
4. Compute the distance r_{max} between the new center point and the vertex furthest from it. If this is less than the previous value of r_{max}, discard the previous center point, keep the current center point, and return to step 2. Otherwise, the previous center point is optimum and the procedure is complete.

This approach is similar to that employed by the Hopfield artificial neural network (Hopfield and Tank, 1985). First, an energy function is defined, in this case the distance from the center point to the furthest vertex. Next, an update rule is also defined, which in this case is the procedure for moving the center point. Finally, the update rule is applied until no further changes are made, and the energy function is minimized.

Once the center point is determined, the conversion of each vertex point (x,y) to polar coordinates (r,θ) takes place as follows:

$$r = \sqrt{(x-x_c)^2 + (y-y_c)^2} \tag{9.19}$$

$$\theta = \tau_\theta \left[\pi + \arctan\left(\frac{-(y-y_c)}{(x-x_c)} \right) \right] \tag{9.20}$$

where $-\pi \leqslant \arctan(a) \leqslant \pi$, and τ_θ is a scale factor. In this implementation of CAMERA, a scale factor of $\tau_\theta = 64/2\pi$ was used to limit θ to 64 discrete values. Because this transform is conducted around the center of the entity rather than some arbitrary point, the resulting representation is shift invariant. During the polar transformation process, the maximum value (r_{max}) of r for all of the vertices in the entity is retained, and then the r values for all of the vertices in the entity are normalized as

$$r = \tau_r \frac{r}{r_{max}} \tag{9.21}$$

where τ_r is the scale factor. In this implementation of CAMERA, a scale factor of $\tau_r = 64$ was used to limit r to 64 discrete values. Because of this normalization, the resulting representation is scale invariant.

Once an entity has been normalized, it is digitized into a 64×64 grid $G(r,\theta)$ by plotting the webs in the entity with r as the vertical axis of the

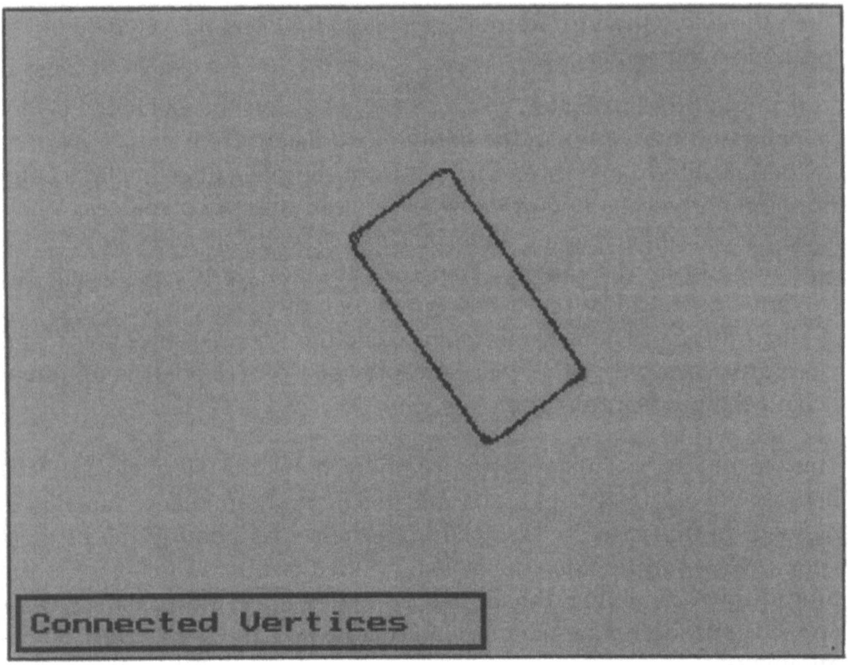

Figure 9.31 Web diagram of a rectangular entity.

grid and θ as the horizontal axis. Although the line segments representing the connections between vertices should be transformed into curves in the polar grid, they are actually plotted simply as straight lines. Because these segments are guaranteed to be relatively short due to the local nature of the vertex connection procedure, the distortion introduced by this simplification is minimal. Figure 9.31 shows an entity representing a rectangle in the original x–y coordinates, and Fig. 9.32 shows the digitized grid that results from the transformation of this entity into normalized polar coordinates.

At this point in the process, the entity representation is invariant to shift and scale, but not to rotation. In order to provide rotation invariance, the grid is shifted in the θ dimension until a balancing function B is maximized. When the shifting is carried out, the columns shifted off of one end of the grid are 'wrapped around' and shifted onto the other end of the grid, so the grid is treated in a circular fashion. The balancing function used is

$$B = \sum_{\theta=0}^{63} r_\theta^* \, b_\theta \tag{9.22}$$

where $r^*_\theta = max\{r|G(r,\theta) = 1\}$. The use of the r^*_θ computation assures that only the extreme outline of the entity is considered in the balancing process. The balancing function can be computed using a neural architecture as shown in Fig. 9.33. The balance weights b_θ are fixed, and

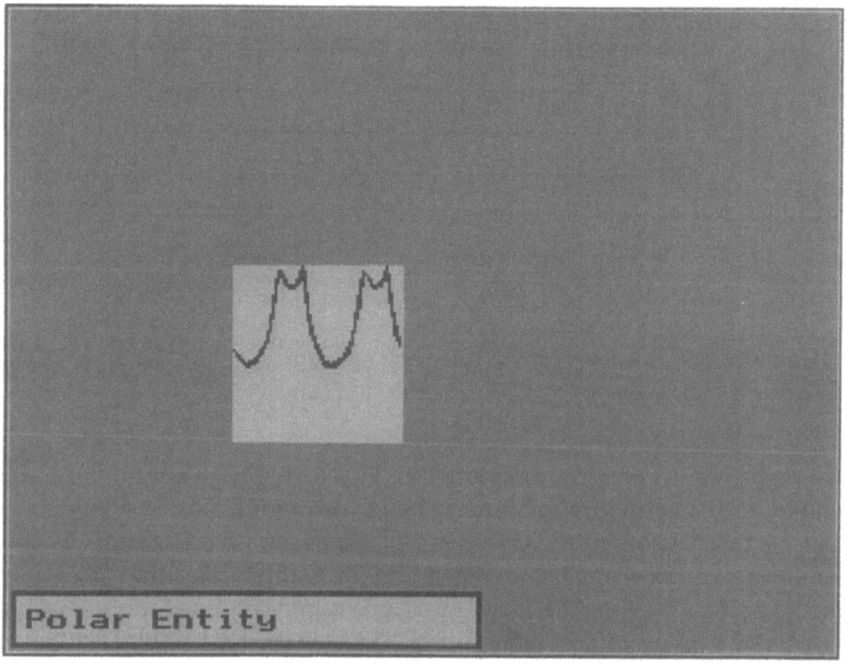

Figure 9.32 Rectangular entity transformed into normalized polar coordinates.

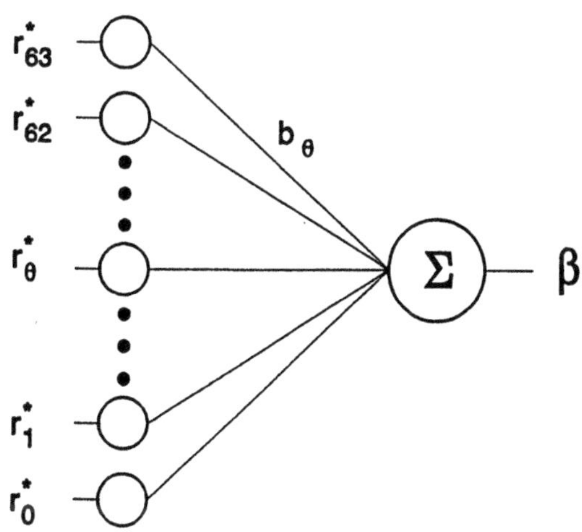

Figure 9.33 Architecture of the balancing operator.

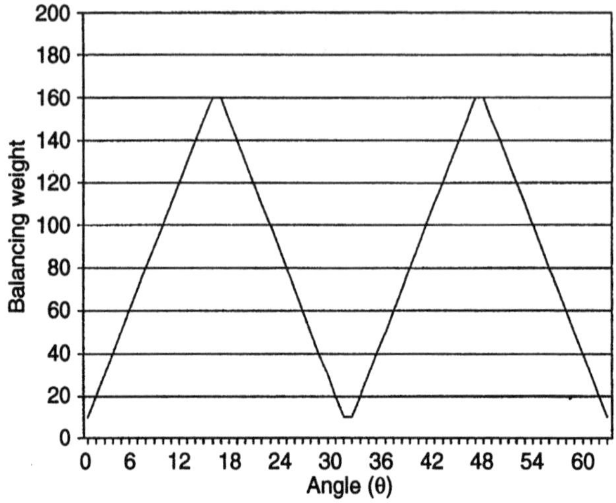

Figure 9.34 Weights used by the balancing operator.

the values of the weights are plotted in Fig. 9.34. Once the grid has been shifted into a configuration where the balancing function is maximized, the representation is *rotation invariant*. Because of the nature of the balancing weights, the balancing process actually locates the major axis of the entity. This results in two possible representations for an entity, one being rotated 180° from the other. This fact is compensated for in the recognition process, which is described next. Figure 9.35 shows the grid of Fig. 9.32 after balancing. The transformed entity, which is shift, scale and rotation invariant, and which is now represented by a 64 × 64 binary grid, is passed to the higher vision stage of CAMERA.

9.4.4 Higher vision stage

The higher vision stage of CAMERA, which performs pattern learning and recognition, consists of the HAVNET artificial neural network. The HAVNET was described in detail in section 4.7, and it was implemented in the CAMERA model exactly as it was described there. The network was designed to accept 64 × 64 input patterns, which were the result of the processing of previous CAMERA stages. Because these representations are oriented along the major axis, the recognition process is actually carried out twice for each input pattern, once in its original form and once with the representation shifted 180°. Because the HAVNET can represent classes of objects as multiple aspects, the CAMERA model is capable of learning to recognize objects not only regardless of scale, rotation and translation, but also under varying conditions. It can also

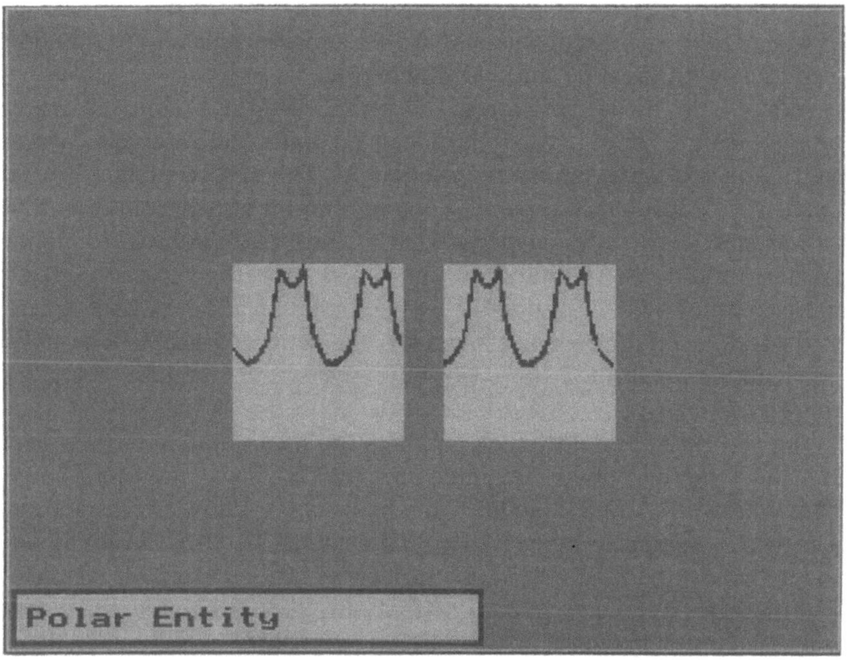

Figure 9.35 Rectangular entity transformed (left) and balanced (right).

learn to recognize three-dimensional objects from several different viewpoints.

9.4.5 CAMERA model summary

In this section, a new artificial vision model (CAMERA) has been introduced. CAMERA has some obvious similarities to the models described previously. First, CAMERA has a structure, first formalized by Marr, in which image information flows from the bottom up, and in which that information becomes more complex but less voluminous at each successive stage. All artificial vision models, and biological vision systems for that matter, share this structure to some degree. Overall, CAMERA is most similar to the Seibert–Waxman model. Edge detection, extraction of vertices, invariant transformation, the use of aspects, and the use of a neural network for object learning and recognition are some of the features that the two have in common. CAMERA has borrowed ideas from other models as well. The webs formed in CAMERA are similar to the boundary webs of the Grossberg model, perceptual grouping is used as in SCERPO, and vertices are used as points at which objects can be separated as in PARVO. Finally, the HAVNET neural network that serves as the highest layer in CAMERA is similar in structure to the neocognitron, and shares some learning and recognition attributes with the ART and SOM networks.

Along with these similarities, however, CAMERA contains many original features, and two developments are particularly notable. First is the connected-vertex entity representation. This representation can be efficiently computed in a parallel manner, and it is rich in information extracted from the original image. It also lends itself very well to higher level processing, since entities can be joined at nearby vertices to create more complex objects, or divided by segmentation at T-junction vertices to separate overlapping objects. Finally, the entity representation can be efficiently transformed into a shift, scale and rotation invariant representation.

The second important development is the incorporation of the novel HAVNET neural network architecture. HAVNET was the first known neural network paradigm to take advantage of the Hausdorff distance as a metric of similarity between two-dimensional patterns. In doing so, the network inherits the desirable properties of the Hausdorff distance, and therefore duplicates human performance more accurately than most previous neural network architectures. The multiple-aspect capability of HAVNET also enables the CAMERA vision model to perform robustly at a wide variety of two-dimensional and three-dimensional object recognition tasks.

Finally, it is worth mentioning that, although biological plausibility was considered during the design of CAMERA, more emphasis was placed on practical implementation of the model because it was intended from the start to implement CAMERA as a functioning artificial vision system. The experimental evaluation of that system is the subject of the next chapter.

10

Experimental evaluation of the CAMERA vision model

10.1 INTRODUCTION

The subject of this chapter is a series of experimental tests that were designed to evaluate the CAMERA vision model. Specifically, the evaluation was intended to determine the extent to which an artificial vision system based on the model met the design objectives stated at the beginning of section 9.4. To briefly review the objectives, they stated that the vision system should be capable of being implemented for a reasonable cost, that it should be capable of recognizing relatively complex objects under a variety of conditions, and that it should recognize individual objects in under 5 seconds. The tests were also intended to test the ability of the HAVNET neural network, which is an integral part of the model, to learn and organize object representations. The series of tests, which are described in the following sections, involved the recognition of numerical digits printed black-on-white, puzzle pieces which represent states taken from a wooden puzzle of the United States, and three common three-dimensional objects, a cup, a flashlight and a wrench.

10.2 EXPERIMENTAL APPARATUS AND CONDITIONS

All of the tests were conducted in the Computer Integrated Manufacturing Laboratory in the Department of Engineering Management at the University of Missouri-Rolla. The images were acquired using an apparatus consisting of a Sanyo model VDC-3800 CCD video camera mounted on a swing arm which was in turn mounted on a table. This arrangement resulted in two degrees of freedom, since the camera arm could swing through over 90° of motion around a horizontal (x) axis, and objects placed on the table could be rotated through 360° about the

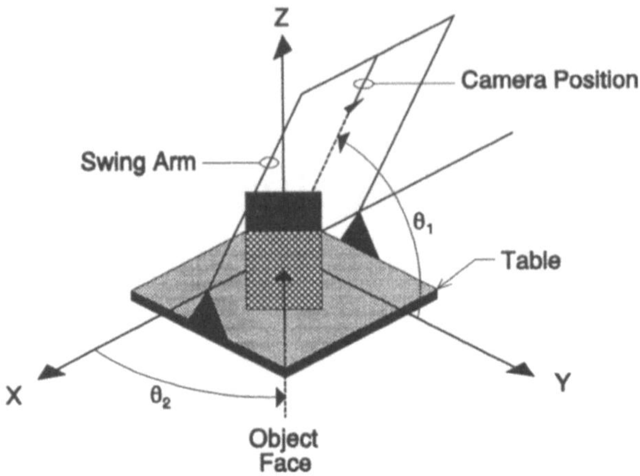

Figure 10.1 Coordinate system of the image acquisition apparatus.

vertical (z) axis. The apparatus was designed such that the vertical and horizontal axes could be made to intersect at approximately the center of the object.

The convention that was used to assign a coordinate system to the apparatus is shown in Fig. 10.1. The angle θ_1 was called the **elevation** angle. It was measured in a right-hand direction around the x-axis, and it varied from 0° with the camera looking directly at the face of the object to 90° with the camera directy over the top of the object. The angle θ_2 was called the **rotation** angle. It was measured in the right-hand direction around the z-axis, and it varied from 0° with the object facing left along the positive x-axis to 180° with the object facing right along the negative x-axis. The side of the object designated as the front was chosen for convenience. The geometry of the apparatus resulted in a distance of approximately 45 cm from the focal plane of the camera to the center of the object, and the camera had a field of view of approximately 24 cm wide by 18 cm high.

The image acquisition apparatus was exposed to the ambient lighting conditions of the laboratory, which consisted of harsh overhead lighting supplied by industrial mercury-vapor fixtures. The apparatus was also equipped with a small incandescent auxiliary light which was mounted on the swing arm next to the camera. This light was used only occasionally, to illuminate the surface features of particularly dark objects. The vast majority of the tests were conducted without auxiliary lighting.

The video signal from the camera was connected to a Data Translation image acquisition (IA) card, model DT55-60. This card was in turn installed in an IBM compatible personal computer with a 386SX- 12 MHz processor. In addition to the computer's standard monitor, additional monitors were used to display the live video signal from the camera and the captured image resident on the IA card. Figure 10.2 is a schematic diagram of this system. The software which implemented the CAMERA vision model and HAVNET neural network, written in Borland Turbo C++, was installed on the personal computer, and the expanded multi-aspect version of the neural network was used in all cases.

10.3 RECOGNITION OF SIMPLE TWO-DIMENSIONAL OBJECTS

10.3.1 Input patterns

The first series of tests were conducted using numerical digits. The digits 0–9 were printed on white paper using a laser printer, with one digit on each page. The digits were treated as objects rather than as printed text. These were simple objects from the standpoint that they were two-dimensional black objects on a white background. Digits represent complex objects, however, from the standpoint that they are not necessarily polygonal, they are not necessarily symmetrical, and they have holes and appendages that can make edge detection and grouping difficult. An illustration of the digits used in these tests is given in Fig. 10.3. All of the images used in these tests were acquired with the camera elevated to 90°, that is, directly over the table. The pages with the digits printed on them were simply placed on the table and imaged, with no auxiliary lighting.

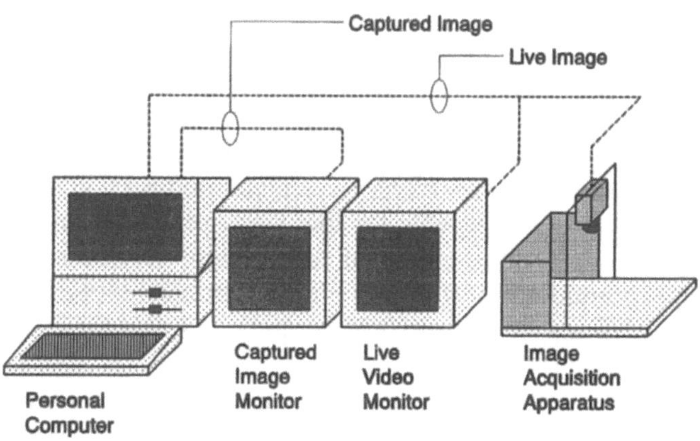

Figure 10.2 Schematic diagram of the CAMERA system implementation.

1 2 3 4 5

6 7 8 9 0

Figure 10.3 Digits used to test learning and recognition of simple two-dimensional objects.

Figure 10.4 Typical training image of a digit.

10.3.2 Object learning

For the digit recognition task the neural network was trained in two passes. Each digit was presented to the system individually, the system was informed of the correct node name for each, and training was conducted. A learning rate of 0.5 was used, and a vigilance parameter of $\varrho = 50$ was used, so a new aspect was generated for a given node only if none of the existing aspects for that node produced an output of at least 50. In the first training pass, all of the digits were presented in their usual top-up orientation, they were located in random positions on the page, and all digits were approximately the same size. Figure 10.4 shows a typical training image.

In the second training pass, the same digits were again presented but with random orientations and positions. Only the digit 'one' generated an additional aspect in the second pass. This was due to a difference in the location of the center of the entity in the two different orientations, resulting in two different polar representations.

10.3.3 Object recognition testing

To test the recognition capabilities of the system a new set of digits was printed, approximately twice the size of that used in training. These digits were presented one at a time in random locations and orientations, and the CAMERA system was asked to recognize the object in the image. Figure 10.5 shows a typical test image. During recognition the numerical output of each network node was recorded, as well as the final decision of the network as to the identity of the object. In order to indicate the identity of the object or objects in the image the CAMERA vision model generates a representation of the input image on the personal computer monitor, marks each object found by surrounding it with its spanning circle, and labels the center of each object with the name associated with the node of the neural network that has identified that object. One such image is shown in Fig. 10.6, illustrating the result of one of the digit recognition tests. The numerical results of the digit recognition tests are presented in Table 10.1.

10.3.4 Discussion of results

The first thing that the results in Table 10.1 demonstrated was that the vision system based on the CAMERA model is capable of learning objects

Figure 10.5 Typical test image of a digit.

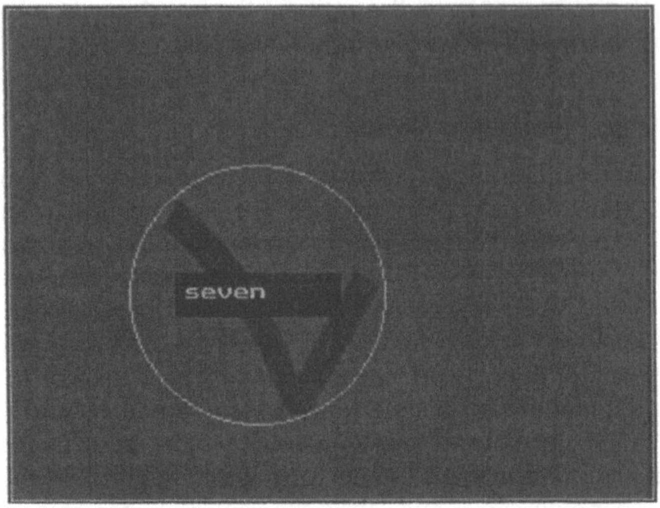

Figure 10.6 Typical result of a digit recognition test.

Table 10.1 Numerical results of digit recognition tests

Input digit	*Recognized digit*	*Neural network output node* 1	2	3	4	5	6	7	8	9	0
1	1	83	28	29	38	27	24	60	26	27	20
2	2	20	73	43	30	27	37	34	43	37	31
3	3	18	30	63	27	30	37	21	46	40	35
4	4	27	31	33	73	32	35	35	35	39	34
5	5	26	37	39	30	63	33	21	37	44	31
6	6	23	30	51	34	47	56	20	52	54	40
7	7	42	40	29	42	24	29	69	30	31	21
8	8	20	37	53	37	36	44	20	60	49	38
9	9	21	33	52	33	44	49	18	50	67	45
0	0	22	29	46	30	24	46	16	42	49	71

from examples and later recognizing them. These results verified the proper operation of the various stages of the model, and demonstrated the system's ability to recognize objects irrespective of scale, rotation and shift in position. Shift invariance was demonstrated because test objects were placed at random positions in the image, scale invariance was demonstrated because the test objects were significantly larger than the training objects, and rotation invariance was demonstrated because test objects were placed at random rotations without adverse impact on recognition results. Rotation invariance was also demonstrated in another manner. The 6 and 9 digits in this font are identical except for rotation. Because of this, the response of both the 'six' node and the 'nine' node were very close when presented with these two patterns. In fact, these two nodes could be considered two aspects of the same object. The fact that each of these digits was recognized correctly rather than confused with the other was most likely accidental. Black objects on a white background were deliberately used in these experiments so that these basic issues of invariant recognition could be tested without the complications of difficulties in edge detection or grouping.

The learning procedure generated two different aspects for the digit 'one', and further investigation revealed that this was due to an inconsistency in the location of the center of the entity prior to polar transformation. A small shift in the center of an entity in rectangular coordinates can change the resulting polar entity rather drastically, which results in the need for an additional aspect to represent the object in question. In order to explain this shift in the position of the center, the center finding process must be reviewed briefly. Recall that, at this point in the processing, the entity is a series of vertices interconnected by short line segments. When an entity is long and thin, as is the case with the digit 'one', the diameter of the minimum spanning circle would typically be defined by the two vertices that are furthest from each other. If all of the processing is ideal this diameter will connect opposite corners of the upright element of the 'one', as shown in Fig. 10.7. The choice of which pair of opposite corners is connected is immaterial, since the center of the minimum spanning circle is what is important, and either choice will result in nearly the exact same center point being

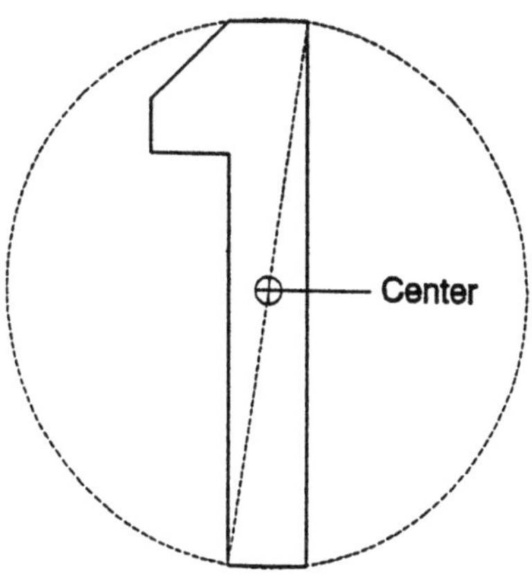

Figure 10.7 Correct centering of the digit 'one'.

generated. The problem occurs when the processing of the image is not ideal. It is possible for a corner of the object to be slightly rounded off in the vertex connection process, and it is also possible for non-linearities in the camera optics or irregularities in the camera's $x{:}y$ aspect ratio to skew the object representation. If either of these occur in the case of an object such as the digit 'one', the diameter of the minimum spanning circle could be erroneously constructed along one of the vertical sides of the upright element of the digit, which would result in the shifting of the center point to the perimeter of that element, rather than the center. Figure 10.8 illustrates this problem. This is in fact exactly what happened in one of the training instances of the digit 'one'. Further testing revealed that this problem was most likely to occur when the digit was presented to the system in an orientation which placed it approximately 45° from either the vertical or horizontal. Instances where the digit was presented either vertically (right-side up or upside down) or horizontally (left or right) always resulted in proper centering.

Another interesting observation that can be made from the above results is that the recognition performance of the system closely resembled the performance of human beings. Table 10.2 was compiled to illustrate this further. This table shows, for each test pattern presented to the system, which nodes were highest, second highest and third highest in output. The table illustrates the fact that the recognition system made the same associations that humans would be likely to. For example, 6 is similar to 9 and 8, 3 is similar to 8 and 9, and 1 is similar to 7 and 4.

It is also interesting to note that the desirable properties of identity, symmetry and triangle inequality exhibited by the Hausdorff distance metric did in fact carry over into the CAMERA vision model. The property of identity was verified by the correct recognition of all digits,

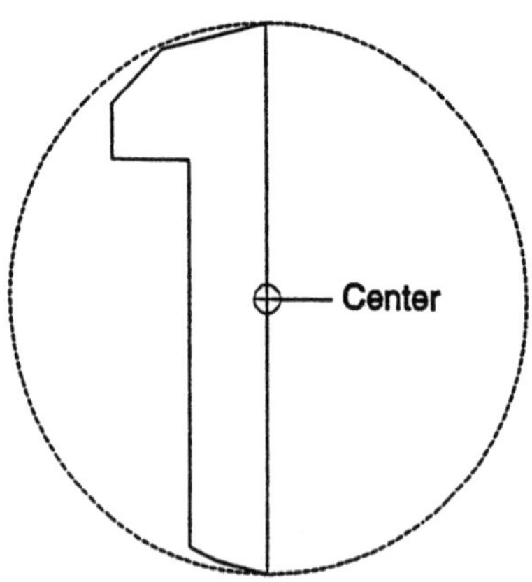

Figure 10.8 Shifted centering of the digit 'one'.

Table 10.2 Similarity between digit objects

Input pattern	*1*	*2*	*3*	*4*	*5*	*6*	*7*	*8*	*9*	*0*
First highest	1	2	3	4	5	6	7	8	9	0
Second highest	7	3	8	9	3	9	1	3	3	9
Third highest	4	8	9	7	8	8	4	9	8	6

demonstrating that an object looks like itself. The property of symmetry states that if object A is similar to object B, then object B should be similar to object A. This was verified, for example, in the case of the digits 1 and 7. The 1 digit scored well with node seven, indicating a high degree of similarity, and the 7 digit scored well with node one, indicating symmetry. The symmetry property was similarly verified with the 3–8 and 6–9 pairs. Finally, the property of triangle inequality states that if two objects are dissimilar, they should not both be similar to a third object. This property was demonstrated, for example, by the digits 0 and 1. Each of these two patterns scored poorly when presented to the other's node, indicating dissimilarity. Furthermore, in no case did both of these patterns score well with any one node, verifying the presence of the property of triangle inequality.

10.4 RECOGNITION OF COMPLEX TWO-DIMENSIONAL OBJECTS

10.4.1 Input patterns

The next series of tests were conducted using puzzle pieces representing states taken from a wooden puzzle of the United States (Milton-Bradley, 1988). These were complex two-dimensional objects in several ways. First, the pieces were of various light colors including purple, green, yellow, orange and pink, making consistent edge detection difficult. Second, the pieces were the shapes of the states represented, so they had complex irregular outlines. Also, like the digits, the states are not necessarily polygonal or symmetrical. In addition, the puzzle pieces had complex surface markings, which included the state name, the locations of natural resources, the names and locations of major cities, and representations of rivers and major lakes. Finally, the states of Colorado and Wyoming were included in the tests because they were essentially identical in outline and differed only in surface detail. The 15 states used in the tests were as follows:

Arizona	Arkansas	Colorado	Idaho	Iowa
Minnesota	Missouri	Montana	Nebraska	Nevada
New York	Oklahoma	Oregon	Texas	Wyoming

All of the puzzle pieces used in the tests are shown in Fig. 10.9. The puzzle pieces were imaged against a light blue felt background, and all of the images used in these tests were again acquired with the camera elevated to 90°, directly over the table. The objects were simply placed on the table and images were taken with no auxiliary lighting.

10.4.2 Object learning

For the state recognition task the neural network was trained in two passes. Each state was presented to the system individually, the system was informed of the correct node name for each, and training was conducted. A learning rate of 0.5 was used, and a vigilance parameter of $\varrho = 60$ was used, so a new aspect was generated for a given node only if none of the existing aspects for that node produced an output of at least 60. In the first training pass all of the states were presented in their usual top-up orientation and they were located in approximately the center of the image. Figure 10.10 shows a typical training image.

In the second training pass the same puzzle pieces were again presented but with random orientations and positions. The puzzle pieces representing the states of Arkansas, Idaho, Montana and Nevada each generated an additional aspect in the second pass. As with the digits, this again appeared to be due to a difference in the location of the center of the entity in the two different orientations, resulting in two different polar representations.

10.4.3 Object recognition testing

To test the recognition capabilities of the system, each of the 15 puzzle pieces was again presented to the system with random positions and orientations, and CAMERA was asked to recognize the object in the image. Figure 10.11 shows a typical test image. During recognition, the numerical output of each network node was recorded, as well as the final decision of the network as to the identity of the object. A typical recognition image generated by the CAMERA system in response to the presentation of a puzzle piece is shown in Fig. 10.12. As in the case of the digits, the image shows the extent of the object and its identification as determined by the neural network. The numerical results of the state puzzle-piece recognition tests are presented in Table 10.3.

10.4.4 Results

The results of the state recognition tests confirmed many of the results of the digit recognition tests. First, the vision system was obviously capable of once again learning objects from experience and later recognizing them. The properties of invariance to rotation and shift in

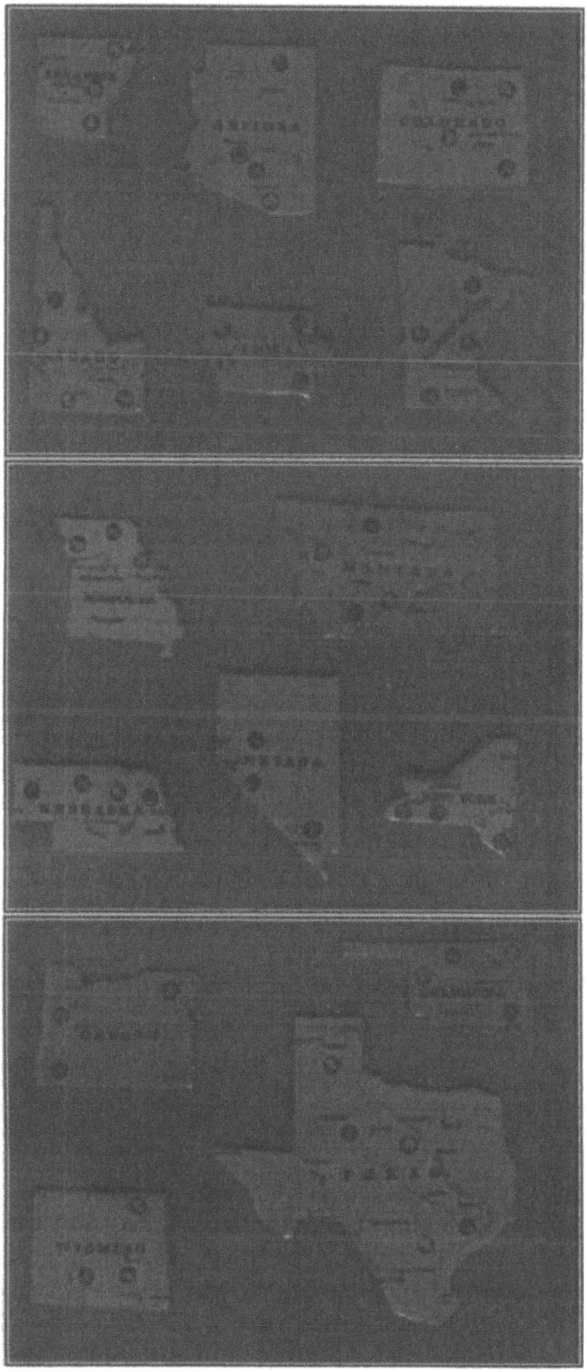

Figure 10.9 Puzzle pieces used to test learning and recognition of complex two-dimensional objects.

Figure 10.10 Typical training image of a state puzzle piece.

Figure 10.11 Typical test image of a state puzzle piece.

position were verified again as well, since the test images involved objects in random rotations and locations. The state puzzle-pieces, however, are far more complex objects than the black-on-white digits. The irregular outlines, color variations and complex surface features present on these objects presented significant challenges to the image processing, grouping, center-finding and balancing stages of the CAMERA vision model. The results of the state recognition test verified that these stages of the model were indeed functioning properly.

Figure 10.12 Typical result of a state recognition test.

Table 10.3 Numerical results of state puzzle-piece recognition tests

Input pattern	*Recognized pattern*	*Neural network output node*														
		AZ	*AR*	*CO*	*ID*	*IA*	*MN*	*MO*	*MT*	*NE*	*NV*	*NY*	*OK*	*OR*	*TX*	*WY*
AZ	AZ	65	56	50	49	49	49	59	37	50	50	52	48	37	52	38
AR	AR	50	66	57	44	49	54	49	40	54	38	59	42	46	49	39
CO	CO	52	55	67	42	52	59	54	43	57	33	54	56	44	48	53
ID	ID	47	40	40	73	41	40	49	47	49	38	40	56	41	54	37
IA	IA	53	59	62	53	87	53	53	44	73	39	54	51	55	51	37
MN	MN	50	53	45	44	37	65	41	41	45	27	47	48	41	48	44
MO	MO	51	48	46	51	46	49	55	49	53	41	51	52	39	49	37
MT	MT	51	48	47	44	46	52	45	81	54	34	44	53	41	48	45
NE	NE	50	55	62	49	64	57	55	66	87	40	48	77	44	48	37
NV	NV	37	39	34	45	35	28	39	36	39	64	37	45	34	43	32
NY	NY	55	62	61	56	53	64	58	42	58	35	76	58	39	53	44
OK	OK	56	49	50	63	48	46	55	48	52	41	49	77	44	47	41
OR	OR	39	35	38	48	36	40	36	34	29	30	34	39	68	39	59
TX	TX	49	44	48	48	41	44	48	40	42	35	40	45	44	55	45
WY	WY	46	41	52	50	37	47	39	33	36	28	35	44	46	54	78

The presence of surface features on these test objects provided an important demonstration of some of the CAMERA vision model's capabilities. Because many of these features were not connected to the outline of the object, they generated separate entities that subsequently had to be grouped properly in order to facilitate correct recognition. In each of the recognition tests, all of the entities present in the image were successfully grouped into a single object representation, which indicated that grouping was occurring properly.

The ability of a vision system to incorporate surface features into the recognition process significantly increases its usefulness. Many industrial applications require that objects with identical outlines be differentiated correctly. For example, many manufacturing processes begin with identical blanks or castings, and later processing involves the machining of different pockets or features into these workpieces or the attachment of small parts that would not change the outline of the piece. In addition, inspection tasks often involve differentiating between objects that differ only in surface appearance. The verification of correct labeling, painting and printing are some examples. A vision system that relied only on object outlines would be useless in these applications. The states of Colorado and Wyoming were included in the test set in order to demonstrate this capability. Although these states had identical outline shapes, the vision system had no trouble differentiating between them on the basis of surface features.

The state puzzle pieces also provided a more challenging test of the recognition capabilities of the HAVNET neural network. The states were much closer to each other in the recognition space of the neural network than the digits were, making them much more difficult to differentiate. In other words, many of the states were very similar to each other, whereas few of the digits were. The puzzle piece representing the state of Missouri provides a good example of this situation. Although the node trained to recognize Missouri was successful in generating the highest output of 55 when presented with the Missouri puzzle-piece, the nodes representing Arizona, Idaho, Nevada, New York and Oklahoma all generated outputs above 50. The fact that many incorrect nodes scored so closely to the correct node is indicative of the difficulty encountered in attempting to differentiate between these patterns.

The learning procedure once again resulted in the generation of an additional aspect for some objects. This appeared to be due to inconsistencies in the center-finding process and in the entity balancing process. Because the balancing process takes place after the center has been found and the entity has undergone polar transformation, inconsistencies in the centering process similar to those described previously for the case of the digit 'one' lead to variations in the behavior of the balancing algorithm. Further testing once again revealed that this problem caused the learning and recognition processes to be somewhat dependent on the rotation of the presented object. Second aspects were most likely to be generated during the learning phase, or matched during the recognition phase, when objects were oriented approximately 45° from either the vertical or horizontal. The causes for these inconsistencies in the centering process appeared to be identical to those described previously for the case of digit recognition.

Like the digit recognition results, the state recognition results once again verified the presence of the properties of identity, symmetry and triangle inequality. Identity was verified by the correct recognition of all states. Symmetry can be observed in the numerical results, since the matrix of numbers that is represented by Table 10.3 should be symmetrical about the main diagonal if symmetry holds. An inspection of the table with this in mind reveals that this is very much the case. The states of Minnesota and Nevada provide a good example of the triangle inequality property. The fact that each of these states scored very low on the node designated for the other indicates that they appear very dissimilar to the network. Further investigation reveals that in no case did both of these patterns score well with any node. In fact, there was no instance in which both of these patterns generated an output of 50 or above when presented to any one node.

10.5 RECOGNITION OF THREE-DIMENSIONAL OBJECTS

10.5.1 Input patterns

The third series of tests was conducted using common three-dimensional objects. These tests had two main purposes: to determine whether three-dimensional objects could be learned and then recognized from characteristic views using aspects, and to determine how the various aspects of an object would be organized by the neural network. The objects used were a cup, a flashlight and a wrench. As with the objects used in previous tests, these objects presented the system with an additional set of challenges. The cup was very dark in color and had little surface detail, but it had a very shiny surface that developed considerable reflections and glare. The flashlight was also dark in color, it possessed a patterned surface texture that made consistent edge detection difficult, and it had a matte finish that made surface features difficult to image. Finally, the wrench had a typical plated metal finish, which caused reflection and glare and caused the surface details of the object to appear significantly different from different camera angles. The three objects used in the tests are shown in Fig. 10.13.

The three-dimensional objects were imaged against a cream-colored felt background, and the image acquisition apparatus was used to collect images of the objects over a range of values for the object rotation and camera elevation angles. Auxiliary lighting was occasionally necessary to prevent the edge representations of the objects from being fragmented beyond the capability of the grouping algorithm to reconstruct the

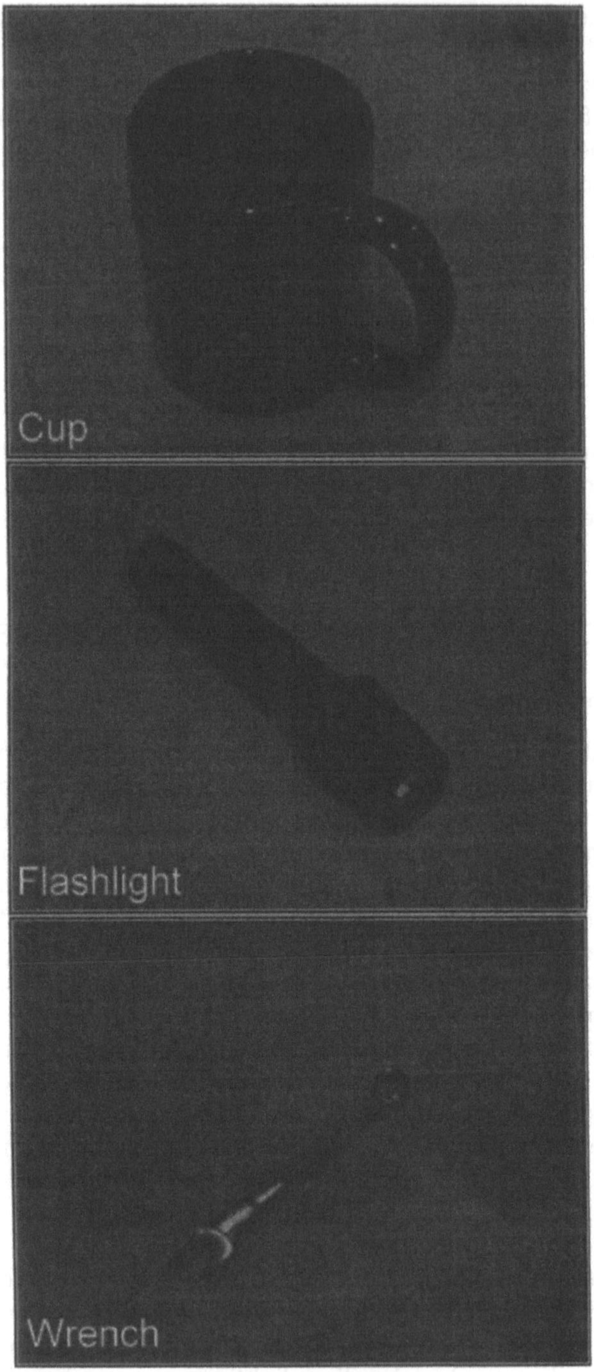

Figure 10.13 The objects used to test learning and recognition of three-dimensional objects.

object. This situation occurred at low elevation angles with the cup and flashlight objects.

10.5.2 Object learning

For the three-dimensional object recognition task the neural network was trained in a single pass. Each object was presented to the system a total of 15 times, under the viewing conditions shown in Table 10.4. The conventions used to indicate the object rotation angles required that one view of the object be designated as the front. For the cup this was the view looking directly at the handle, for the flashlight this was the view looking directly into the lens, and for the wrench this was the view looking directly into the open end.

The neural network was trained to learn all 15 of the views for each of the three-dimensional objects. A learning rate of 0.5 was used, and a vigilance parameter of $\varrho = 50$ was used, so a new aspect was generated for a given node only if none of the existing aspects for that node produced an output of at least 50. During the training procedure, the camera swing arm was positioned at each of the elevation angles, and each object was imaged at each of the rotation angles. In all cases the elevation and rotation angles were approximated, and were not the result of rigorous measurements. Figure 10.14 shows a typical training image for the flashlight object.

Table 10.4 Viewing angles used for three-dimensional object training

Camera elevation (θ_1)	*Object rotation* (θ_2)
0	0
0	45
0	90
0	135
0	180
45	0
45	45
45	90
45	135
45	180
90	0
90	45
90	90
90	135
90	180

Figure 10.14 Typical training image of the flashlight object.

One of the objectives of this experiment was to determine how the neural network would self-organize the various aspects or characteristic views of a three-dimensional object into a coherent object representation. Figure 10.15 shows how the 15 views of the cup object were organized by the neural network into eight nodes. The same sort of analysis was conducted with the flashlight and wrench objects. Figure 10.16 shows how the 15 views of the flashlight were organized by the neural network into four nodes, and Fig. 10.17 shows how the 15 views of the wrench object were organized into four nodes. Representing all 45 of the learned views of these three objects, then, required a total of 16 neural network nodes.

10.5.3 Object recognition testing

To test the three-dimensional recognition capabilities of the CAMERA vision model, each of the three objects were presented to the trained vision system under approximately the same viewing conditions that were used during training, i.e. those listed previously in Table 10.4. Again, the angles were approximated and not measured rigorously. For each test the selected object was placed on the image acquisition table in the proper orientation, the camera elevation angle was adjusted as desired, and the vision system was asked to identify the object in the image. Figure 10.18 shows a typical test image. During the recognition process the numerical output of each neural network node was recorded, as well as the final decision of the network as to the identity of the object. For each learned object the number of the aspect generating

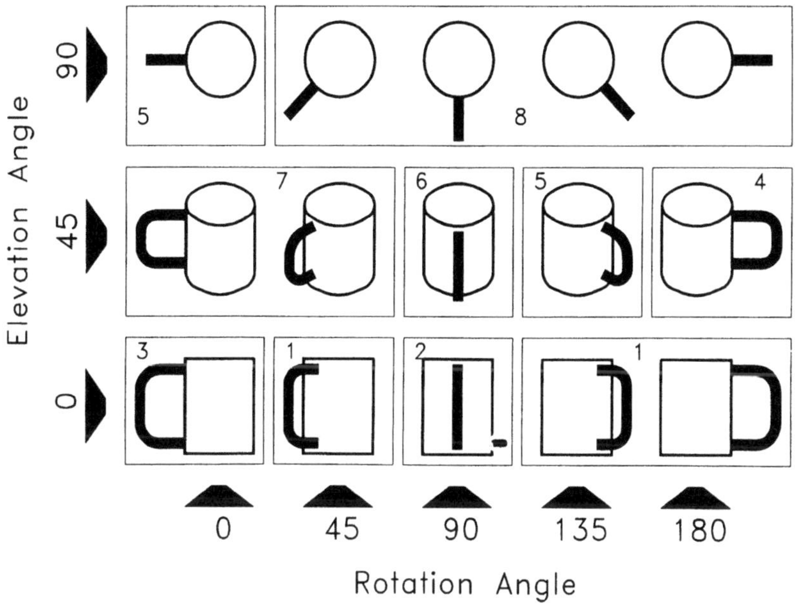

Figure 10.15 How the neural network organized aspects of the cup object during training.

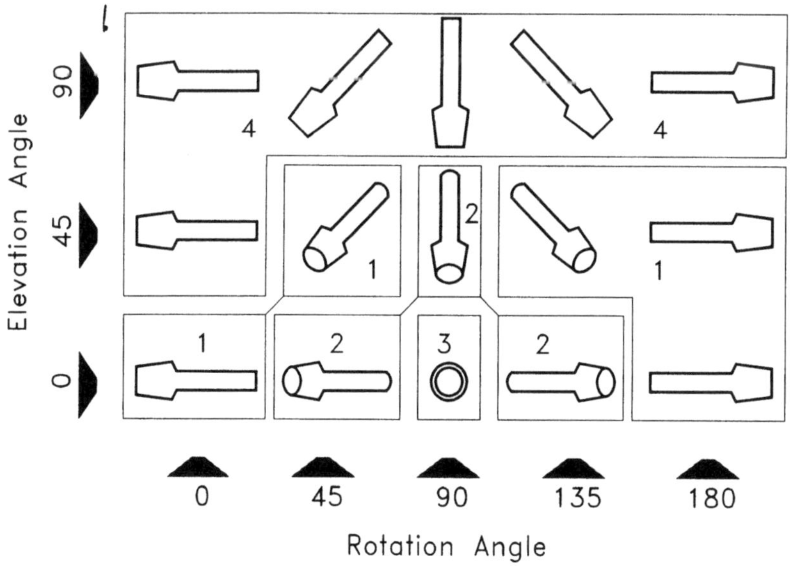

Figure 10.16 How the neural network organized aspects of the flashlight object during training.

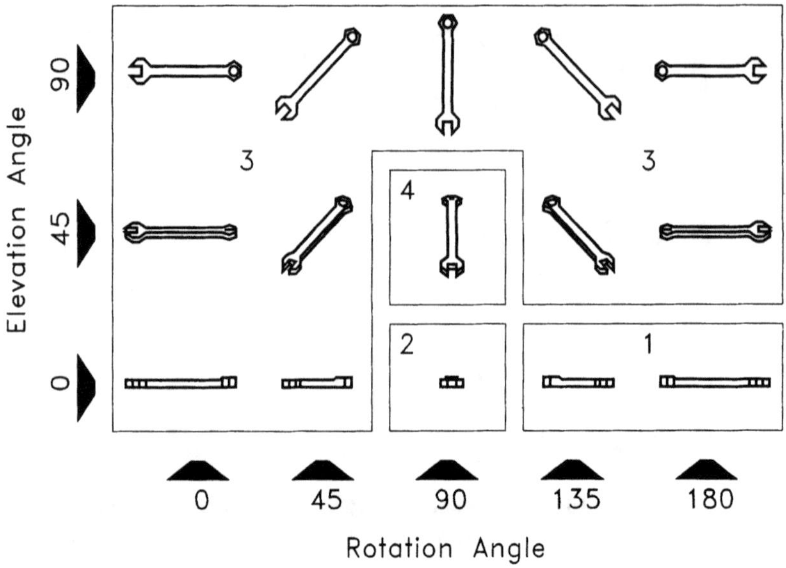

Figure 10.17 How the neural network organized aspcts of the wrench object during training.

Figure 10.18 Typical text image of the wrench object.

the highest output in response to the current input image was also recorded. A typical recognition image generated by the CAMERA system in response to the presentation of one of the three-dimensional objects is shown in Fig. 10.19. As in previous tests, the image shows the extent of the object and its identification as determined by the neural network.

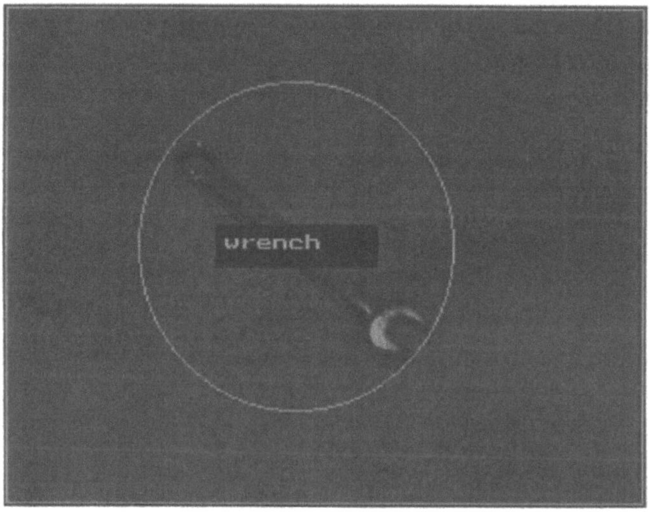

Figure 10.19 Typical result of a three-dimensional object recognition test.

The numerical results of the three-dimensional object recognition tests performed on the cup, flashlight and wrench are presented in Tables 10.5, 10.6 and 10.7, respectively. Only the outputs of the node representing the highest-scoring aspect for each object are shown in these tables, and the number of the highest-scoring aspect for each learned object is shown to better illustrate the multi-aspect recognition process.

10.5.4 Results

The primary capability demonstrated by the results presented above was the fact that the CAMERA vision system was able to learn three-dimensional objects from multiple views, and later recognize those objects. The system was trained on 15 views of three different objects, and it was later capable of recognizing those objects under similar viewing conditions. This capability would obviously be very useful in a manufacturing or packaging environment. Not only would the general purpose learning and recognition of three-dimensional objects be advantageous, but also the ability to recognize three-dimensional objects from a limited number of views. For example, if parts such as bolts were transported on a conveyor-belt and later identified by a vision system, they could be recognized correctly whether they were standing on end or had fallen down. This gives an additional degree of freedom to material handling systems. The same would hold true for many other parts, as well as bottles, cans, boxes and many other packages.

Table 10.5 Numerical results of cup recognition tests

Presented object	*Elevation* θ_1	*Rotation* θ_2	*Recognized object*	*Aspect matched*	*Cup output/ aspect*	*Flashlight output/ aspect*	*Wrench output/ aspect*
Cup	0	0	cup	3	53/3	32/4	31/2
Cup	0	45	cup	3	41/3	29/3	36/2
Cup	0	90	cup	2	40/2	30/4	28/2
Cup	0	135	cup	3	46/3	27/3	37/4
Cup	0	180	cup	3	55/3	31/2	26/4
Cup	45	0	cup	7	51/7	40/4	36/2
Cup	45	45	cup	7	50/7	39/4	44/2
Cup	45	90	cup	6	43/6	24/4	29/3
Cup	45	135	cup	5	51/5	41/4	38/2
Cup	45	180	cup	4	86/4	40/4	34/3
Cup	90	0	cup	5	60/5	58/4	44/3
Cup	90	45	cup	8	51/8	37/4	37/2
Cup	90	90	cup	8	55/8	35/4	31/3
Cup	90	135	cup	8	46/8	37/4	32/2
Cup	90	180	cup	8	86/8	42/4	34/3

Table 10.6 Numerical results of flashlight recognition tests

Presented object	*Elevation* θ_1	*Rotation* θ_2	*Recognized object*	*Aspect matched*	*Cup output/ aspect*	*Flashlight output/ aspect*	*Wrench output/ aspect*
Flashlight	0	0	flashlight	4	26/8	80/4	31/3
Flashlight	0	45	flashlight	2	27/4	68/2	36/2
Flashlight	0	90	flashlight	3	37/1	46/3	36/4
Flashlight	0	135	flashlight	2	31/4	68/2	37/2
Flashlight	0	180	flashlight	4	34/8	75/4	42/3
Flashlight	45	0	flashlight	4	26/8	67/4	27/2
Flashlight	45	45	flashlight	4	30/8	72/4	43/2
Flashlight	45	90	flashlight	4	28/8	64/4	35/2
Flashlight	45	135	flashlight	4	25/3	61/4	32/2
Flashlight	45	180	flashlight	4	26/8	63/4	25/2
Flashlight	90	0	flashlight	4	25/8	69/4	26/2
Flashlight	90	45	flashlight	4	26/8	76/4	27/2
Flashlight	90	90	flashlight	4	22/4	61/4	28/2
Flashlight	90	135	flashlight	4	27/8	72/4	28/2
Flashlight	90	180	flashlight	4	26/8	61/4	24/2

Table 10.7 Numerical results of wrench recognition tests

Presented object	*Elevation* θ_1	*Rotation* θ_2	*Recognized object*	*Aspect matched*	*Cup output/ aspect*	*Flashlight output/ aspect*	*Wrench output/ aspect*
Wrench	0	0	wrench	1	36/8	29/1	65/1
Wrench	0	45	wrench	3	27/6	30/1	54/3
Wrench	0	90	wrench	2	29/2	19/3	42/2
Wrench	0	135	wrench	3	31/6	30/1	56/3
Wrench	0	180	wrench	1	42/6	35/1	75/1
Wrench	45	0	wrench	3	42/6	46/1	90/3
Wrench	45	45	wrench	3	44/6	54/4	73/3
Wrench	45	90	wrench	4	22/4	23/4	31/4
Wrench	45	135	wrench	3	42/6	56/4	87/3
Wrench	45	180	wrench	3	38/6	44/1	82/3
Wrench	90	0	wrench	3	45/6	51/1	99/3
Wrench	90	45	wrench	3	44/6	53/4	92/3
Wrench	90	90	wrench	3	44/6	55/4	92/3
Wrench	90	135	wrench	3	45/6	53/4	99/3
Wrench	90	180	wrench	3	42/6	61/1	99/3

The characteristics of the three-dimensional test objects also provided additional evidence of the robustness of the processing stages of the CAMERA vision model. Of all of the test objects used in the research to this point, these three were the most difficult to edge detect and group consistently. As was the case during training, the darker cup and flashlight objects required the use of auxiliary lighting at low camera elevation angles in order to produce an image that resulted in correct edge detection and grouping. This appeared to be due to the need to reduce the effect of the shadows cast by the harsh overhead ambient lighting. These three objects also presented the CAMERA system with other challenges. It is interesting to note that of all of the views recognized by the system, the view of each object that was most difficult for the system to recognize (based on the strength of the network output) was the direct view of the object face ($\theta_1 = 0°$, $\theta_2 = 90°$). These views are shown in Figs 10.20–10.22. This was another example of the similarity between the CAMERA system performance and human performance. All three of these difficult views would also be the most likely to be confused by humans. The view of the cup does not show the distinctive handle feature clearly, and the views of the flashlight and wrench also provide little detail that is useful for recognition purposes. Another indication of the difficulty in recognizing these views is the fact

Figure 10.20 Difficult view of the cup object.

Figure 10.21 Difficult view of the flashlight object.

that for each of these views a separate aspect was generated, and that aspect was not used later to recognize any other view of the object.

Another purpose of this series of three-dimensional recognition tests was to determine how the learned aspects generated by the neural network were later used during recognition. The way in which the

Figure 10.22 Difficult view of the wrench object.

learned aspects were utilized during recognition for the cup object is shown pictorially in Fig. 10.23. Note that although eight aspects were generated during the cup learning process, only seven of those were used during recognition. Cup aspect 1 was unused. A similar diagram for the flashlight object is shown in Fig. 10.24. Again one aspect, flashlight aspect number 1, was not used during recognition. A similar diagram was also developed for the wrench object and it is shown in Fig. 10.25. In the case of the wrench, all of the learned aspects were used during recognition. The reason that some learned aspects are later not used in the recognition process appears to stem from the fact that an aspect better representative of the characteristic view that generated the original aspect is generated later in the learning procedure. In each case where aspects were not used, the views that originally generated those aspects were recognized by aspects generated later in the learning process. It is possible that the neural network nodes representing these unused aspects could be deleted without adversely affecting recognition performance.

10.6 RECOGNITION ACCURACY

10.6.1 Purpose

The purpose of the next series of tests was twofold. First, it was desired to perform a large number of recognition tests in order to determine the

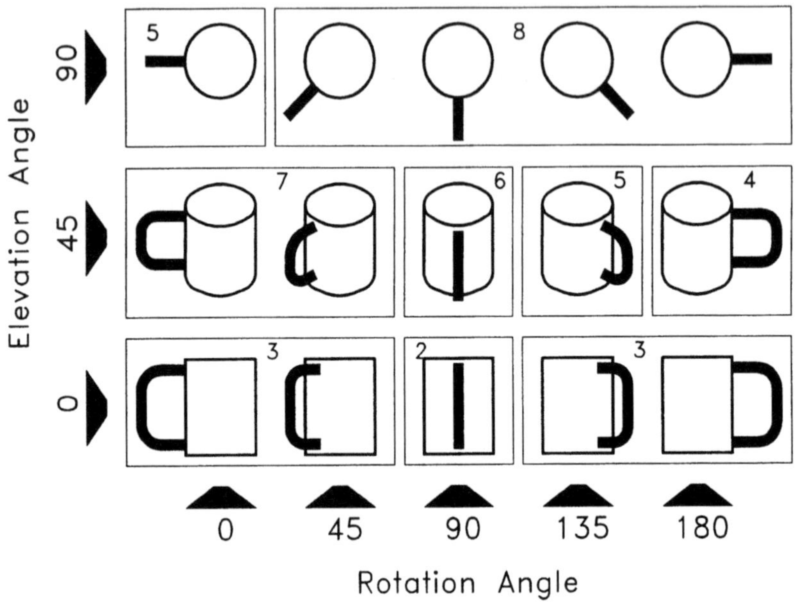

Figure 10.23 Learned aspects used during recognition of the cup object.

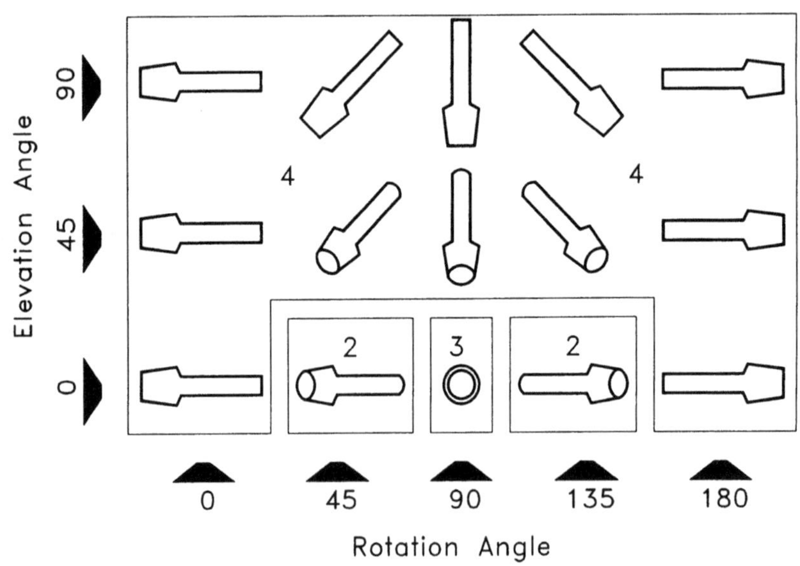

Figure 10.24 Learned aspects used during recognition of the flashlight object.

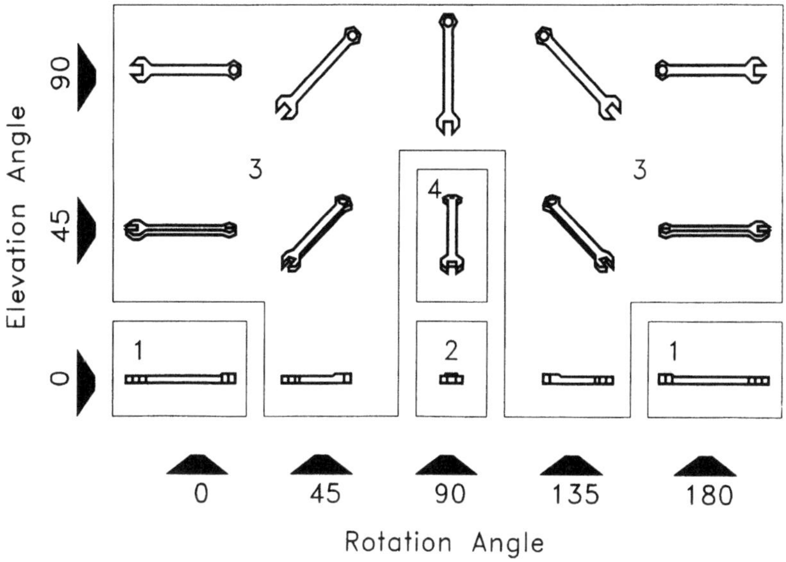

Figure 10.25 Learned aspects used during recognition of the wrench object.

overall recognition accuracy of the CAMERA vision system. Second, it was desired to test the capability of the system to reject as 'unknown' objects which it had not been trained to recognize.

10.6.2 Test procedure

For the recognition accuracy tests the 15 state puzzle pieces were again used. The neural network weights that were the result of the previous training on the states were again used, so no additional training was conducted. Twenty groups of five states each were randomly generated in such a way that each state appeared in at least six groups and no state appeared in more than seven groups. The puzzle pieces representing these states were then presented to the vision system in the same manner as was done previously, with the exception that now all five states in any group were presented to the system in a single image. The puzzle pieces were arranged in random positions and orientations, although they were required to be fully visible in the image and a minimum separation of approximately 1 cm was maintained between all pieces. Figure 10.26 shows a typical test image containing five pieces.

Once presented with the image, the vision system was asked to identify all of the objects in the image, and the identification results were recorded. Figure 10.27 shows the recognition results generated by the CAMERA vision system in response to the image of Fig. 10.26. The

Figure 10.26 Typical test image for recognition accuracy tests.

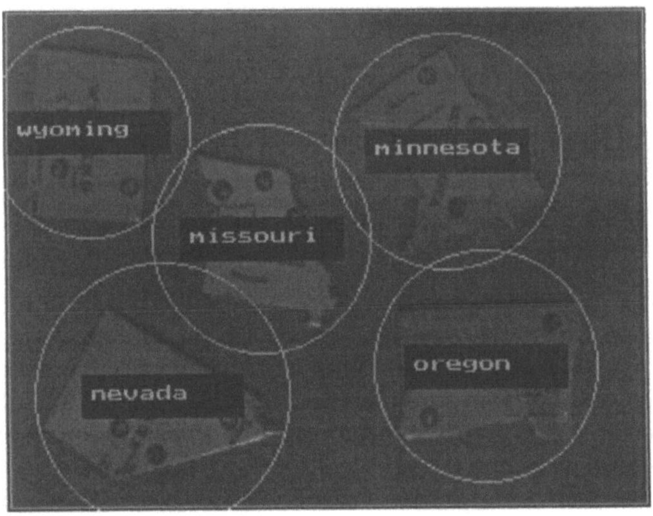

Figure 10.27 Typical result of a recognition accuracy test.

numerical results of all of the recognition accuracy tests are presented in Table 10.8. Note that, of a total of 100 individual recognitions performed during these tests, 96 resulted in correct identifications, yielding an overall recognition accuracy of 96%.

The rejection tests were performed with the same neural network as

Table 10.8 Results of recognition accuracy tests

Group number	*States presented*	*States recognized*	*Success ratio*
1	OK TX IA OR WY	ID TX IA OR WY	4:5
2	MN AR MT MO AZ	MN AR MT MO AZ	5:5
3	NV NY NE ID CO	NV NY NE ID CO	5:5
4	CO NE MT AR TX	CO NE MT AR TX	5:5
5	NY OK IA AZ ID	MN OK IA AZ ID	4:5
6	NV MO MN OR WY	NV MO MN OR WY	5:5
7	NE AR NY MN TX	NE AR NY MN TX	5:5
8	ID OR NV MO CO	ID OR NV MO CO	5:5
9	AZ MT OK WY IA	AZ MT OK WY IA	5:5
10	AR NY OR NE OK	AR ID OR NE OK	4:5
11	NV ID CO IA TX	NV ID CO IA TX	5:5
12	MO AZ MN MT WY	MO AZ MN MT WY	5:5
13	MN AR MO CO WY	MN AR MO CO WY	5:5
14	OK TX ID NY MT	OK TX ID NY MT	5:5
15	NV AZ NE OR IA	NV AZ NE OR IA	5:5
16	NY WY MN NE IA	NY WY MN NE IA	5:5
17	AZ ID NV CO MO	AZ ID NV CO MO	5:5
18	AR MT OK TX OR	AR MT OK TX OR	5:5
19	MN MO OK CO AR	NY MO OK CO AR	4:5
20	NE NY NV AZ OR	NE NY NV AZ OR	5:5

was used in the tests described above, so again no training was involved. These tests involved presenting the trained vision system with puzzle pieces representing six states that the network had not been trained to identify. These pieces represented the states of North Carolina, Ohio, South Dakota, Utah, Washington and Wisconsin, and they are shown in Fig. 10.28. Each of these was presented to the system individually, and the magnitude of the output from each neural network node was recorded. The numerical results of the rejection tests are presented in Table 10.9.

10.6.3 Results

The recognition accuracy results demonstrated two important capabilities of the CAMERA vision system. First, its ability to analyze an image with multiple objects, segment the information in that image correctly into object representations, and recognize each of the objects present regardless of position or orientation. Second, these objects were recognized with a high degree of accuracy.

Figure 10.28 Puzzle pieces used in the rejection tests.

Table 10.9 Numerical results of state puzzle-piece rejection tests

		Neural network output node														
Input pattern	*Recognized pattern*	*AZ*	*AR*	*CO*	*ID*	*IA*	*MN*	*MO*	*MT*	*NE*	*NV*	*NY*	*OK*	*OR*	*TX*	*WY*
NC	MT	49	50	45	49	45	56	57	58	57	46	51	54	38	43	39
OH	NY	52	62	66	52	56	52	61	51	65	37	69	51	44	48	40
SD	MT	49	55	63	50	63	56	55	63	56	41	46	40	52	59	42
UT	OR	51	40	36	46	36	41	43	48	37	41	39	39	55	48	50
WA	AR	49	63	56	49	56	58	55	58	65	39	62	47	43	47	43
WI	MO	53	50	50	47	50	54	60	54	52	47	46	48	42	55	43

Of the four cases in which the states were incorrectly identified, three of them involved the New York puzzle piece. In two cases New York was improperly identified, and in another case Minnesota was identified as New York. Further testing revealed that, because the state of New York is approximately triangular in shape, the balancing process was inconsistent. This could result in up to three different choices of major axis, and each of these would have to be represented by a different aspect. As luck would have it, the one aspect that was generated for New York was also similar enough to the representation of Minnesota to cause an error.

The other incorrect case involved the misidentification of the state of Oklahoma as Idaho. Inconsistent centering appeared to be the cause of this error, and additional training would probably have generated another aspect for Oklahoma to compensate for this. The fact that Idaho had already generated two aspects in the neural network made its identification more likely. In short, most of the errors experienced in these tests could probably be eliminated with additional training. Even without additional training, however, the system recorded a very favorable 96% rate of correct recognition.

In order to evaluate the results of the rejection tests a quantity called percentage confidence (P_c) was defined. This quantity is intended to compare the level of output of the winning node in the neural network with the average outputs of all of the other nodes. Percentage confidence was computed using the equation

$$P_c = \frac{\text{net} - \text{net}_{\text{ave}}}{\text{net}} \times 100 \tag{10.1}$$

where net is the output of the winning node, and net_{ave} is the average output of all non-winning nodes. Percentage confidence values based on the numerical results of the rejection tests are presented in Table 10.10. Percentage confidence values were also computed for the original state recognition test results. These computations yielded figures ranging from 14% for Missouri to 46% for Wyoming, with an average of 33%.

A threshold level could be set for these confidence values, so that when the value falls below the threshold the object could be considered unknown. There are two approaches to selecting this threshold, one where false identifications are costly and must be avoided and another where a good guess is better than nothing. With the first approach a confidence threshold of 25% could be applied to the test results described here. With this threshold level all of the untrained states would result in responses of 'unknown'; however, the states of Missouri (14%) and Texas (20%) that were originally identified correctly would also be classified as 'unknown'. With the second approach a confidence threshold of 20% could be used. This would result in Texas being identified correctly, Missouri being classified as 'unknown', and with the identities of the unlearned states of Ohio and Utah being 'guessed at' as New York and Oregon, respectively.

Table 10.10 Percentage confidence for rejection tests

Pattern presented	*NC*	*OH*	*SD*	*UT*	*WA*	*WI*
Pattern recognized	MT	NY	MT	OR	AR	MO
Percentage confidence P_c	16	23	18	22	17	18

10.7 RECOGNITION TIME

10.7.1 Procedure

In order to determine the total time required for the CAMERA vision system to recognize objects from images, a version of the program that implements the vision system was created that was capable of reading input images from files rather than from the image acquisition card, and which computed and displayed total elapsed time. This was done so that recognition times could be conveniently measured with the program running on a wide variety of computers. Elapsed time was begun immediately after the image was loaded into memory from a file, and before any image processing. Elapsed time was stopped and displayed when all of the objects in the image had been identified.

To provide inputs for the program, two images were captured and saved to disk files. One contained the single digit 'five', while the other contained five state puzzle pieces (AZ, MO, NE, NY and OR). These image files were copied to a diskette along with the modified program and the neural network weight files for the previously trained digit and state recognition networks. The program was then executed on several different computers using the identical procedure. The results of the timed digit and state recognition on the various computers are shown in Table 10.11.

10.7.2 Results

Because only one or two of these computers were actually equipped with the image acquisition card, these times do not include image acquisition and compression time. These activities required less than 0.5 s on the Pentium-based (fastest) computer. With image acquisition and compression time excluded, the recognition of a single simple object required 1.48 s on the fastest computer. The recognition of five fairly complex objects required 7.08 s on the same computer. With image

Table 10.11 Numerical results of timed recognition tests

Computer	*Single-digit recognition time (s)*	*Five-state recognition time (s)*
386SX 12 MHz	38.34	136.44
386DX 16 MHz	29.08	101.23
386DX 33 MHz	13.35	49.54
486DX 33 MHz	7.08	30.87
486DX 66 MHz	3.51	15.11
Pentium 90 MHz	1.48	7.08

acquisition and compression time included, the five complex objects were recognized in approximately 7.5 s, an average of 1.5 s per object present.

10.8 SUMMARY OF CAMERA EVALUATION

10.8.1 Experimental conclusions

The proper operation of the various stages of the CAMERA vision model was verified by the test results just presented. The Sobel edge detection operator performed very well, and the edge enhancement algorithm did an excellent job of thinning and defining the edges. The vertex extraction and connection processes performed very well over a wide variety of conditions, and the entity grouping and invariant transformation stages also performed correctly in diverse situations. The model proved capable of generating object representations that were invariant of shift, rotation and scale. These representations were compact and yet rich in information about the original object. The system was also capable of grouping entities into multiple objects, and presenting each of those objects for recognition.

The capabilities of the HAVNET neural network, which serves as the upper stage of the model, were also tested. The network was capable of learning and later recognizing, with a high degree of accuracy, a wide variety of objects under diverse conditions. The recognition accuracy of printed digits presented individually was 100%, the recognition accuracy of state puzzle pieces presented individually was 100%, the recognition accuracy of three-dimensional objects (a cup, a flashlight and a wrench) presented individually was 100%, and the recognition accuracy of state puzzle pieces presented in many randomly selected groups of five was 96%. In addition to the recognition tests, the ability of the network to reject unknown objects was tested. It was found that all unknown objects could be rejected successfully at the expense of also rejecting some previously learned objects. The aspect generation and organization capabilities of the network were also tested. Aspects were generated and organized adequately to allow for the correct recognition of three-dimensional objects from 15 different viewpoints, and the number of aspects generated was significantly fewer than the number of views. Some of these aspects were not used during recognition, indicating that some views were over-represented in the neural network.

10.8.2 Assessment of the model

One of the reasons for conducting the experimental evaluation was to determine to what extent a vision system based on the CAMERA model met the design objectives set forth at the beginning of this study. The first objective stated that it should be possible to implement the vision system for a reasonable cost using existing technology. The vision system as tested here could be implemented for a total cost of less than $5000, including a video camera, an image acquisition card, and a Pentium 90 MHz personal computer. The second objective stated that the system should be capable of correctly recognizing reasonably dissimilar objects under various conditions. Two-dimensional and three-dimensional objects of a variety of shapes, sizes, materials, textures and colors were all recognized with a high degree of accuracy against backgrounds of various colors and under differing lighting conditions. Also, objects that were identical in outline and which were somewhat similar in appearance were differentiated correctly, based solely on the appearance of surface features. The final objective stated that the system should be capable of recognizing modestly complex objects in under 5 s. The state puzzle pieces, which had complex outlines, were multicolored, and which had complex surface markings, were recognized in an average time of 1.5 s per object present.

Some secondary design guidelines were also set forth at the beginning of this study. These stated that the resulting vision system should be capable of learning objects from examples, that it should employ a high degree of parallel processing, and that it should be as biologically faithful as possible. The application of biologically inspired neural-like processing in the lower levels of the model and the incorporation of an artificial neural network as the object recognition mechanism represent design attributes that helped meet these goals. The capability, speed, and robustness of the final design provide testimony to the wisdom of employing these design guidelines from the start.

Another desire expressed at the outset of this study was that of developing a vision system that would be capable of performing a wide variety of tasks, as opposed to one that was designed for a specific application. The fact that the identical vision system performed all of the tasks in this study, from two-dimensional black-on-white digit recognition to three-dimensional object recognition, is evidence of its flexibility. The fact that the system was capable of generating multiple aspects to represent objects was in part responsible for this flexibility. Aside from the obvious purpose of storing different views of three-dimensional objects, aspects provide a mechanism for the system to compensate for a host of practical difficulties, including image distortions, inconsistencies in processing, shadows, glare and changes in lighting conditions. A system without multiple aspect capability simply would not be capable of performing at the level that was demonstrated in this study, even in the case of two-dimensional object recognition.

The overall robustness of the CAMERA model was also evident in the consistency of the level of recognition performance over a wide variety of objects. Although the recognition tests involved sets of objects that were progressively more difficult to learn, recognize and differentiate, the recognition accuracy of the system did not deteriorate as the more challenging tests were conducted.

The ability of the system to handle multiple objects in a single image also improved its robustness. Familiar objects in a scene could be recognized even in the presence of unknown objects, or even entities generated as artifacts of the image acquisition or processing procedures. This capability makes the CAMERA system considerably more useful than systems that are capable of only single-object recognition. Also contributing to robustness was the fact that the processing employed in CAMERA resulted in object representations that were invariant of shift, scale and rotation. This invariance allowed many views of a particular object to be consolidated into a small number of aspects, and it enabled the system to generalize between trained views. The wrench provided a good example of this, where 11 different views were recognized by one neural network node representing a single aspect of that object.

One aspect of the CAMERA vision model that continually surfaced during the experimental evaluation was the similarity between its performance and human visual tendencies. The CAMERA system had difficulty with certain views of three-dimensional objects, and these same views were those most difficult to recognize by humans. Also, when the CAMERA model detected similarities between objects, it was likely that those same objects would seem similar to human observers. Particularly encouraging was the fact that the CAMERA model demonstrated human like adaptability and robustness under varying conditions. The fact that the CAMERA model employed biologically inspired processing may have been responsible for these similarities.

10.8.3 Potential applications

A general purpose vision system capable of learning from examples, such as the CAMERA system described here, would obviously have many applications in manufacturing environments. The ability to learn and later recognize two-dimensional and three-dimensional objects is a human capability that is employed on many assembly and processing lines, and a machine-based system that could perform these tasks would have many benefits. Two applications, namely material sorting and visual defect detection, are described next as representative examples.

Material sorting is a difficult and complex task to perform automatically. Whether the job is to sort manufacturing workpieces or recyclable materials, the problems are very similar. A system capable of performing this task would have to be capable of identifying a wide variety of three-dimensional objects from various views, and of locating those objects for automatic picking. The orientation of the object would also be important in many cases. The CAMERA vision model could be employed to perform such a job. The system could be trained to recognize the objects from multiple views, and it could then be employed on-line to locate and recognize the objects. Because information on the position of the major axis of an object could be preserved during the balancing process employed by CAMERA, the object's orientation could be communicated together with location and identity information to a material picking system like a robotic manipulator.

The capability of the CAMERA system to identify and locate multiple objects in a single image would make it particularly applicable to this situation, since objects to be located and identified would not have to be presented to the system individually. An automated system for the task of material sorting could potentially be cheaper, faster and more accurate than human operators, and it could operate constantly in the undesirable conditions present in many manufacturing plants and recycling facilities.

Visual identification and classification of defects is another difficult task normally performed by humans. Two product lines that could benefit from the automation of this task are sheet goods (for example paper, metal, fabrics and photographic film) and semiconductor materials. All of these products are typically inspected for visible defects, and it is often desirable to locate the defects and classify them into categories. The CAMERA system would be well suited to this task. The system could be trained to recognize defects from examples, and neural network nodes could represent defect categories. The percentage confidence measure could be employed to sort unrecognized defects into an 'other' category. Images could be taken of the surface of the material in question, and the defects in those images could be identified, classified and located. Statistical information on the frequency of each of the defects could be generated, as well as maps indicating defect locations. Based on this information, decisions could be made as to the suitability of the material for further processing.

The ability of the CAMERA model to employ multiple aspects would be particularly useful in this application, as defects that are classified into a single category often differ somewhat in visual appearance. As in the previous example, an automated system for this task could potentially be cheaper, faster and more accurate than human inspectors, and it could operate constantly without suffering from fatigue. Also, the potential for increased inspection speed with an automated system could enable 100% inspection of some of these products, where the limited inspection of random samples is now the norm. This would have the advantages of both preventing defective product from shipping and rejecting as unsuitable only product which is known to be defective.

10.8.4 Recommendations for future research

Two classes of recommendations for future research are elaborated upon in the following paragraphs. The first group addresses weaknesses in the CAMERA vision model as it is currently specified. These are deficiencies that became apparent during the experimental evaluation of the model. The second group of topics includes features that could be added to the CAMERA model specification, in order to make its performance even more human-like.

Current model deficiencies

The primary weakness in the implementation of the CAMERA vision model that was employed in this study was the fact that the segmentation portion of the perceptual grouping and segmentation stage of the model was not operational. Without this segmentation capability the system was not capable of separating objects that were touching or overlapping. Segmentation capability would also allow the system to deal with objects imaged against a cluttered background, whereas the present implementation of CAMERA requires a relatively clean background behind the object. The decision was made to exclude the segmentation algorithm and to concentrate on the other aspects of the model when it became obvious that a considerable amount of research and development effort would be required to perfect the segmentation procedure. The interconnected vertex object representation lends itself well to segmentation, but the identification of segmentation points and the practical implementation of procedures to subdivide the entities involved proved to be very complex. Because of this, preliminary work has been done towards using stereo (depth) information as a segmentation aid. Additional details of this work are given in the following section.

One other deficiency that became evident during testing was the inconsistent behavior of the entity centering procedure. Possible causes for this problem are non-linearities in camera optics and difficulties in edge detection and vertex finding. Further research could lead to more consistent behavior of these processing steps, or to some non-linear transformation that could be applied to an image prior to further processing.

Another problem encountered in the tests was the fact that, in order to guarantee complete rejection of unknown objects, confidence threshold levels would have to be chosen that would also result in the rejection of some known objects. Additional training could possibly solve this problem, since the competitive aspect of the training algorithm had very little effect with the small number of training passes employed during these tests. Additional training would result in the similarities between objects being ignored while their distinctive features would be emphasized. Some preliminary work has been done in this area with very promising results. A more powerful competitive learning algorithm was implemented in the HAVNET neural network, and the network was used to recognize handwritten digits (Rosandich, 1995). The application of competitive learning on this task improved recognition accuracy from 58% to 91%.

Another possible improvement in the way that the HAVNET network is currently implemented in CAMERA would be the elimination of learned aspects that are not utilized during recognition. This feature would improve the efficiency of the recognition process and reduce the size of the neural network required. One possible approach to this problem would be to implement a form of slow degradation in the neural network weights, a degradation that would be reversed for any node each time that node was used to recognize an object. In this way nodes that lie unused for long periods of time would eventually have their weights reduced to very low values. A simple check of weight values for each node could then be used to delete the unused nodes.

Processing speed is another possible drawback of the current CAMERA model implementation. Although the recognition time was well below the 5 s target, many practical applications require even shorter processing times. The parallel nature of many of the processing steps employed by the model makes its implementation on a parallel processor a natural next step. An excellent computer for this task is the CNAPS neurocomputer by Adaptive Solutions, Inc. The CNAPS parallel neurocomputer is a single instruction multiple data (SIMD) parallel processor. It has 256 processors, each a pipelined unit optimized for fixed point multiply/accumulate calculations. Each processor has 4 kbytes of local memory, and the architecture supports up to 64 Mbytes of global memory. The architecture is particularly suitable for image processing, pattern recognition and neural network tasks. A preliminary study was conducted as part of this research to help determine the improvement in recognition time that could be expected by implementing the CAMERA vision model on this computer. The results of that study indicated that recognition times of less than 0.2 s per object present would be achievable. An excellent topic of future research would be the implementation of CAMERA on this computer or one with similar capabilities.

Potential additions to the model

Color information is a powerful discriminant in object recognition. This is evident from the use of color coding on many products and in many other everyday situations. The addition of color capability to CAMERA would allow the model to take advantage of this fact to improve the accuracy of object recognition. For example, objects that are virtually identical except for color could be easily discriminated by a color capability, whereas they would probably be confused by an edge-based recognition system. The addition of color sensing also has the advantage of allowing the system to model human performance more accurately. In tasks like medical diagnosis and product defect analysis, color is often the primary discriminant employed by humans. Color discrimination is also important to such tasks as analyzing aerial or satellite images to determine the type of foliage or ground cover present. Such tasks would be within the capability of an artificial vision system that employed color processing.

Color sensation could be added to the CAMERA vision model by converting the camera and image acquisition board to color (typically RGB). The color information would then have to be incorporated into the object representation if it was to be used in the recognition process. The simplest way to do this is by implementing region filling. The regions segmented by the boundary webs generated by CAMERA could be filled with the appropriate color, which could be the average color of the associated region in the original image. A few modifications would be necessary to implement region filling in the object representation used in CAMERA. After the entity representation is transformed, normalized and balanced, any non-edge pixels in the 64×64 grid would be assigned the color associated with that region. This information would then be used in the recognition process. In addition to the Hausdorff distance, which represents the degree of simplicity between the edge locations, a color-distance metric would be employed to indicate the similarity of color between the input pattern and the learned representation. The neural network architecture would be modified to include color learning as well as edge learning.

One of the difficulties imposed by the introduction of color processing is the additional complexity of the vision system. Color images require more memory, and the associated processing is more complex. The region-filling and object-recognition operations would also add to system complexity. Another problem stems from the fact that the vertex connection process employed by CAMERA is not flawless, and partially bounded regions would complicate the region filling process. One other possible problem is that color processing will complicate the edge detection process. Also, the addition of color sensing is sure to increase the overall cost of the resulting system, as color cameras and image acquisition boards are significantly more expensive than their grey-scale counterparts. Finally, the additional processing steps required for color vision are sure to increase processing and recognition time.

Another possible improvement to the CAMERA vision model is the addition of stereo vision. The introduction of stereo vision to the model could improve the segmentation and grouping performance of the intermediate vision stage. Webs of the same depth often represent components of the same object and could be grouped together, and an abrupt depth change in a web could be used to indicate a segmentation point. Depth information has been shown to be very useful in separating overlapping objects.

Some preliminary work has been completed towards the goal of including stereo segmentation into the CAMERA model. A dual-camera arrangement similar to that explained in Chapter 7 was used, and both left and right images were processed up to the point of vertex extraction and connection. Stereo information was then used to assign depth information to vertices. The edge matching technique explained in Chapter 7 was implemented as a solution to the correspondence problem in order to accomplish this. Common depth was used as a grouping aid, in addition to the containment and proximity measures already in place. More importantly, the depth information was used to segment overlapping objects rather than the method of segmentation at T-junctions. By searching for discontinuities in depth, the webs formed in the grouping process were segmented into representations of separate overlapping objects. This technique proved to be capable of reliably segmenting overlapping objects that were separated by at least 1 cm in depth.

Two problems caused by the introduction of stereo vision to the model are cost and complexity, since the majority of the vision model has to be duplicated. Other problems stem from the computations required to compute depth from stereo and the difficulties associated with solving the correspondence problem (Chapter 7). The edge matching process currently used to determine stereo correspondence is rather simple and does not perform well in all situations, making depth computations prone to error. Finally, the additional processing required for stereo vision increases the processing and recognition time for the model.

The ability to generate camera movements that serve to center an object in the field of view could also improve the capabilities of the CAMERA model. Camera movements could eliminate the need for translation and rotation-invariant processing, so the object recognition process could be simplified. There is ample evidence for this type of behavior in humans, as objects of interest are almost always kept in the center of the field of view. Camera movement also has several other advantages. When ambiguity is present in an image which makes recognition impossible, moving the camera to acquire a slightly different view will often eliminate the ambiguity (Seibert and Waxman, 1992). Camera movement would also allow a system like CAMERA to explore and learn its environment, as the different views required to learn a three-dimensional object would be acquired automatically through appropriate camera movements. Camera movements would also allow the system to deal with moving objects, since movements could be generated that would track an object long enough for it to be recognized. There is evidence of visual tracking in humans and animals. Finally, camera movements could be used to acquire two slightly different views of a scene, and therefore enable the stereo processing described in the previous section with a single camera.

The movement of the human eyes and head correspond roughly to changing the orientation of the camera (eye movement) and changing the position of the camera (head movement). Both of these capabilities could be incorporated into the vision model by mounting the camera on a robot arm. A three-degree-of-freedom (DOF) wrist could be used to orient the camera, and a three-DOF arm would serve to position the camera. Control of the manipulator could be accomplished through serial communication. Further flexibility could be gained by mounting the robot arm on a mobile platform, resulting in a mobile exploratory vision system. Desired moves for the manipulator or mobile platform could be generated based on input from the vision system.

Camera mobility is not without its problems, however. As with previous enhancements, the primary difficulties with this added capability are increased cost and complexity. Robot manipulators and mobile platforms are not inexpensive, and the computational burden involved with generating appropriate signals for position and orientation control of the camera is significant.

The ability to compute or estimate the camera view angle is also important to three-dimensional object recognition. Aspect graphs are used to represent three-dimensional objects in CAMERA, with each aspect representing a characteristic view of an object. If the orientation of the camera were known with respect to the viewed object, the number of candidate aspects could be reduced to only those that are consistent with the known information. If this capability were incorporated as part of a mobile exploratory system, camera orientation information would allow a higher level of world modeling, where a three-dimensional model of the entire environment could be developed rather than individual object representations. This type of model would greatly reduce the number of candidate objects considered by the recognition process at any given position.

Sensing the direction in which the camera is pointing is fairly easy if the robotic camera positioning system described previously is employed. If the base coordinates of the robot arm or mobile platform were known, the camera orientation could be computed based on feedback of position and orientation variables. This technique would minimize the additional complexity required in order to incorporate this feature.

One difficulty introduced by the incorporation of viewpoint sensing is the ill-posed nature of the problem of object location given only the camera orientation. With only this information, an object can be located along a particular line, but the position along that line is unknown. This problem could be solved with the addition of range information from stereo as described previously, or from focus information as described next.

Active focus control could further increase the flexibility of the CAMERA model. The first benefit from active focus control would be the guarantee of high quality images, a fact that would certainly improve the overall performance of the system. Also, the feedback information from the focusing system could provide an indication of depth or range that could be used in much the same way as stereo information to segment or group objects. In addition, the size of an object could be computed from the range information, making it possible to discriminate between objects of different scales that are otherwise identical.

One way to introduce active focus control to the model is to employ an auto-focus camera (with a local focus control loop) in place of the manually focused camera presently in use. The difficulties with this approach are that auto-focus systems are notorious for constantly changing the focus and therefore altering the image at often undesirable times, and that no feedback of focus position is normally supplied. A more predictable approach would be to implement active focus control within the CAMERA vision model. One approach to this problem which uses only image information has been published (Lee *et al.*, 1991). This approach analyses intensity gradients in an image along a line of pixels, either vertical or horizontal or both. A quantity representative of the quality of focus is computed along the line or lines. This quantity gets larger as focus improves, and it has been shown to be independent of illumination levels. This approach could be implemented by adjusting the camera focus to maximize this quantity before capturing an image for analysis. A feedback of camera lens position could also be used as an indication of the approximate range of the object in the image.

One problem with active focus control is the additional expense of the focus control and feedback hardware on the camera. Also, although this active focus control technique appears promising, it may in fact suffer from some of the problems associated with existing auto-focus technology. In addition, the focusing process would have to be carried out before an image was acquired, lengthening the total processing time. Finally, the feedback of lens position is sure to be a highly non-linear and approximate indication of object range, since a relatively small lens movement typically changes the focus from 6 inches to infinity.

Incorporating many of these suggested capabilities into the CAMERA vision model would create a truly powerful artificial vision system capable of more human-like performance. Many of these improvements could interact to help solve the problems of grouping, segmentation, color discrimination, three-dimensional object recognition and environment modeling. With these improvements CAMERA could evolve into an autonomous system that could locate, learn and identify all of the objects in its environment through intelligent exploration.

Part Four

Case Studies

11

Automated visual inspection systems

11.1 INTRODUCTION

It is the objective of this chapter to demonstrate how some of the techniques discussed in this book can be applied to real inspection problems. Two problems are covered in the following sections, the inspection of polished silicon wafers and the inspection of pharmaceutical blister packages. The systems proposed to solve these problems are similar in that they are automated, they are designed to improve on human inspection and on commercially available inspection systems, and they are capable of learning defect categories from examples.

The two systems also have some differences, however. The wafer inspection system represents implicit inspection (Chapter 2) which involves searching for and identifying defects of unknown size, rotation and location on a product. The blister-pack inspection system, however, represents explicit inspection, where the structure of a product is known and can be compared to a template or model. This chapter shows that an intelligent automated visual inspection system based on artificial neural networks can function very well in either of these inspection domains.

11.2 INSPECTION OF POLISHED SILICON WAFERS

11.2.1 Problem description

Polished silicon wafers provide the raw material from which very large scale integrated (VLSI) circuits are manufactured. As VLSI designs become simultaneously more dense and more complicated, the probability of a defect in the wafer affecting the functionality of the final circuit becomes significantly greater. This has caused VLSI manufacturers to demand wafers of very high quality. They also require some indication of the location and density of the defects on a wafer, so that designs can be placed in such a way as to minimize the number of defective circuits.

Driven by these quality demands, a manufacturer of polished silicon wafers for the VLSI market has expressed a desire to automate the wafer inspection processes. Current procedures include inspecting the front (polished) side of the wafer for defects, inspecting the back (unpolished) side of the wafer for defects, and measuring the bulk denuded zone across the thickness of the wafer. The purpose of these inspections is to provide statistical data to customers about defect density and distribution, and to provide feedback about the performance of the wafer manufacturing process. This manufacturer presently does all inspections manually.

Inspections are presently done using binocular microscopes with Nomarski differential interference contrast (DIC) optics, using reflected light. The DIC optics enhance the contrast of scenes that vary in depth but are essentially monochromatic, and this is the case with wafer inspection. Wafer surface defects are nearly invisible under a standard microscope. Prior to inspection, a wafer is etched to reveal the defects. About 2 μm of the surface is etched off, and this step causes the inspection to be destructive. Wafers etched in this manner are useless for further processing.

Manual front-side inspection is typically done at 400× magnification, and inspections are done using one of two patterns, either the *X–Y* pattern or the nine-point pattern. The *X–Y* pattern involves counting and classifying the defects in two continuous strips, one vertical and one horizontal. The strip width is equal to the microscope view-field width at the magnification used. The strips are centered on the wafer, so the exact center area is counted twice. This is the method of counting currently preferred by the manufacturer. With normal defect density, a 6 in. wafer takes about 5–7 min to inspect manually using the *X–Y* pattern.

The nine-point pattern is used when defect density is too high to count manually with the *X–Y* pattern. Defects in nine locations are counted and classified, with each location being one microscope view-field in size. The locations are the center of the wafer, four points on the perimeter (north, south, east and west) and four points at one-half radius out from the center (N, S, E, W).

With either manual technique the results from the pattern inspection are used to estimate the overall density and distribution of defects. Other techniques are used to inspect for back-side defects and to measure the bulk denuded zone of the wafer, and these will not be elaborated upon here.

11.2.2 Front-side inspection feasibility testing

It is the wish of the manufacturer to implement automated inspection in phases beginning with front-side defect detection, since it is believed that automating this phase of the inspection process will have the greatest benefit for the investment. With this in mind, a study was conducted in order to determine the feasibility of locating defects on the front side of the wafer and classifying them into categories. Front-side inspection currently consists of a manual search for defects, with the emphasis being on the identification of two different classes of known defects which will be identified here as defects S and H. All other visible defects are identified as 'other'. The manufacturer wants the automated system to be capable of identifying both the S and H defects, and of learning new defect categories if they should arise.

For the feasibility testing, the manufacturer supplied black-and-white Polaroid photographs of typical defects, taken through the microscope at 400× magnification. These images were then digitized and analyzed. The resulting images represented approximately 20 instances of defect H and 40 instances of defect S. Digitizing the photographs resulted in graininess in the images, so pre-processing was done by applying a smoothing filter.

An artificial vision system based on the CAMERA vision model presented in Chapter 9 was developed to analyze the defect images. This system allowed defects to be classified and located regardless of scale or rotation, and it enabled defect categories to be learned from examples. Images representing approximately 15 instances of defect H and 30 instances of defect S were presented to the system for training purposes, and each defect in the training images was identified to the system during training. Figure 11.1 shows representative examples of H defects, and Fig. 11.2 shows representative examples of S defects. Two passes through the training images were made with a learning rate of 0.50. After training, the system was capable of locating and identifying the defects in the training images with 100% accuracy.

Figure 11.1 Three examples of H defects used to train the inspection system.

Figure 11.2 Three examples of S defects used to train the inspection system.

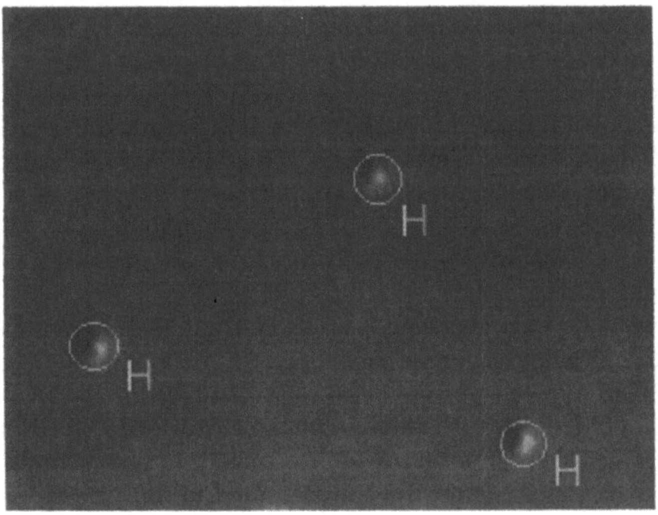

Figure 11.3 Results of H defect identification test.

The remaining images, representing approximately five instances of defect H and 10 instances of defect S, were used to test the system. The images were presented to the system, and it was asked to identify all of the defects present. It is the nature of this problem that both types of defects are not likely to appear in a single image, and that was the case for the test images. Figure 11.3 shows a test image in which all of the H defects have been correctly located and identified, and Fig. 11.4 shows a test image in which all of the S defects have been correctly identified. These figures represent typical results of the defect recognition tests.

Although this represents a very limited test, some general conclusions can be drawn from the results. First, H defects were much more difficult to recognize than S defects. Where results indicated that S defects could be identified with approximately 95% confidence, H defects could only

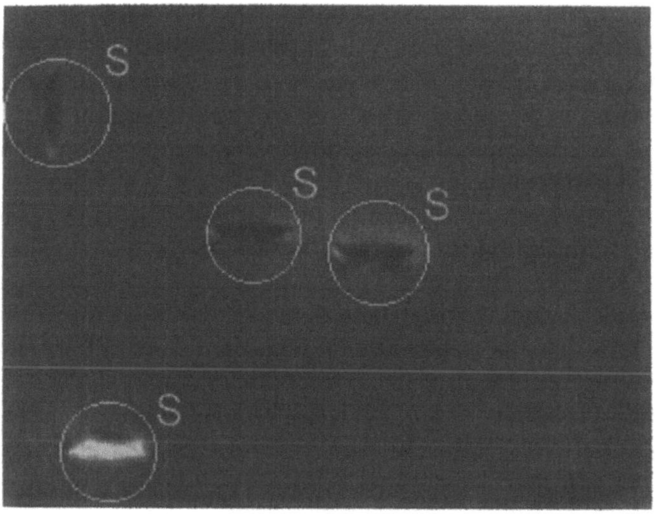

Figure 11.4 Results of S defect identification test.

be identified with about 70% confidence. This is most likely due to two factors. First, the size of the H defects is quite small in the images, so there is little detail for the system to work with in identification. Second, the appearance of H defects is not very distinctive. Whereas the S defects have a distinctive trapezoidal shape, H defects appear as small bumps that cast a fuzzy shadow, making them more difficult to identify. Identification errors, however, only resulted in defects being identified as 'other' rather than H or S. In no instance was a visible defect missed entirely.

Another general conclusion drawn from the test results was that it is feasible to learn defect categories from examples. The two categories used in the tests were learned simply by presenting the system with example images of those defects, and new categories could be learned in the same way. This represents a significant improvement over most automated inspection systems, where the introduction of new defect categories usually involves some reprogramming.

11.2.3 Proposed automated wafer inspection system

The next phase in this project will involve the development and implementation of the final automatic wafer inspection system. It is expected that the final system will be capable of all three inspection types, namely front-side inspection, back-side inspection and bulk denuded zone measurement. The system will also be based on the CAMERA vision model, and as such will be capable of learning new defect categories when provided with example images of those defects. A general description of the operation of the proposed automated wafer inspection system follows.

To set the system up for conducting an inspection the operator will input an identification number for the wafer to be tested. The operator will then select the type of inspection to be conducted (front side, back side or bulk denuded zone). The operator will then input the size of wafer or type of wafer fragment to be inspected (6 in. full wafer, 8 in. full wafer, 1/4 section of 6 in. wafer, 1/4 section of 8 in. wafer, pie section of 6 in. wafer, pie section of 8 in. wafer). The level of magnification and the inspection pattern to be used (nine-point estimate, *X–Y* scan, or 100% inspection) will then be selected.

A microscope will be purchased and fitted with a CCD grey-scale camera. The microscope will be equipped with DIC optics, automated adjustments for magnification and focus, and an *X–Y* table for wafer positioning. The system will require accurate placement of the wafer to be inspected on the *X–Y* table by the operator or by an automatic wafer delivery system. Automatic adjustment of magnification, focus and Nomarski DIC optics will be performed, and the *X–Y* table will be automatically controlled to perform wafer scanning in the desired pattern. Images will be acquired as the wafer is scanned, with each image located in wafer-centered coordinates.

Images will be analyzed immediately on-line by the CAMERA vision model. All defects found will be located in wafer coordinates and classified into established categories, or 'other'. A total raw count of each category of defect will be reported, along with a calculation of defect density (per square centimeter) overall, and for each category of defect. A map of defect density on the wafer and a plot of radial defect density will be generated. All of the above data will be displayed in a graphic format on a single computer screen, with an option to print any data. The data will also be archived to disk as required, saved by wafer ID number.

To adapt the system to new defect categories, the operator will put it into 'learning' mode. A wafer with the defects of interest will be placed on the *X–Y* table, and the location of the defects and magnification to be used will be entered. An image will be acquired and displayed on the screen, where the operator can mark defects of interest with a pointing device, and enter a new defect category name. The images of the marked defects will be stored by the system as examples for training. Once enough examples of a certain defect have been gathered, the system will be trained to recognize this defect in the future.

11.2.4 Wafer inspection system summary

The system proposed in this section represents a novel approach to a challenging automated inspection and defect classification problem. Automating the procedure enables 100% inspection of the surface of wafers selected for testing, rather than the manual pattern-based inspections currently used. This in turn enables much more data to be generated from each wafer inspection, including defect density maps and accurate plots of radial defect density. Based on input from the wafer manufacturer, the quality of the automated inspection would be consistent with results generated by the best human inspectors. The automated system has the additional advantages of being faster (a 100% inspection could be conducted in less time than a manual pattern inspection) and more consistent. Fatigue effects are also eliminated with the automated system. Current manual inspectors are limited to only a few hours of inspection duty per shift due to the concentration required. Inspection cost is also reduced, since a single operator could oversee several automated inspections simultaneously. Finally, because the automated system employs artificial neural network technology, it is capable of learning new defect categories as they arise.

11.3 INSPECTION OF PHARMACEUTICAL BLISTER PACKAGES

11.3.1 Problem description

Many pharmaceutical products, both prescription and over-the-counter, are supplied in blister packages. These packages are usually constructed in two pieces, a blistered container of foil or plastic with a pattern of blisters to contain a certain quantity of pills or capsules, and a cover of foil or plastic film to contain the product. During the packaging process the empty blister packages move through a production line where they are filled with product. The usual practice is to package a single pill or capsule in each blister.

It is desirable to inspect the packages before the cover is applied and sealed, since detecting and correcting incorrectly packaged products at this stage is fairly easy and inexpensive. It is also necessary for many product lines to be 100% inspected as mandated by governments for health and safety reasons. The inspection process involves stopping the package momentarily on the production line and acquiring an image for analysis. The image of the package is typically divided into rectangular cells, with a blister location at the center of each cell. The images of the cells are then compared to some standards to determine if the correct product is present or if some packaging defect has occurred. Typical defects are missing product, broken product, wrong product or multiple product.

11.3.2 Blister pack inspection feasibility testing

A study was conducted to determine the feasibility of the automated visual inspection of blister packages. For purposes of the study, each blister location was to be classified as good product, broken product, missing product or other. Additional defect categories could be introduced to the system at a later time.

Images were acquired of test packages with examples of many good products and a few missing and broken products. The images were acquired under very high-contrast lighting conditions, so the blister package itself was essentially invisible. Only shadows of the edges of the blisters appeared in the images. Because the location of the product in the image is fixed, the image could be divided into cells based on the expected locations of each blister. This avoided any searching and speeded up the inspection process considerably.

As with the previous example, an artificial vision system based on the CAMERA vision model was developed to analyze the images of the product. The system was first trained to recognize good, missing and broken products, and then tested on typical images of filled blister packages. To provide training data, the area of the image corresponding to each blister location was extracted from several training images that contained representative examples of good products and of the defects. A square image area (100 × 100 pixels) centered on the location of each blister was extracted and presented to the inspection system, along with the correct identification of the product (good, missing or broken). Figure 11.5 shows representative examples of a good product, a missing product and a broken product. As in the previous example, two passes through the training images were made with a learning rate of 0.50.

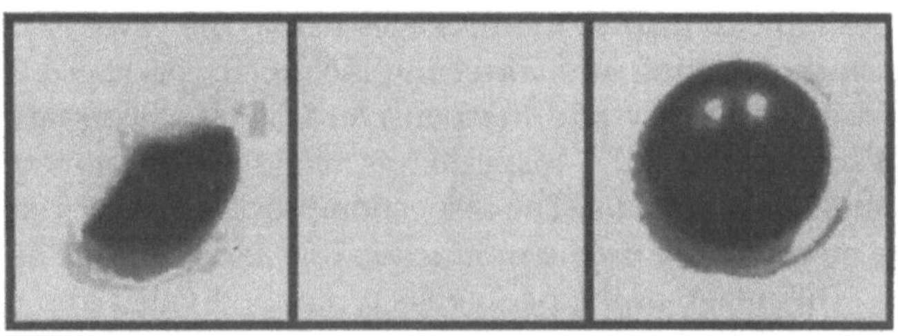

Figure 11.5 Typical training examples for a broken (left), missing (center) and good product.

After training, the system was again capable of identifying the products in the training images with 100% accuracy.

Additional images were acquired for testing purposes. These images represented many examples of good products, with a few broken and missing products present in random locations. The images were presented to the inspection system, and it was asked to classify all of the products present into one of the three established categories, or 'other'. Figure 11.6 shows a typical test image in which those blisters not classified as good have been correctly identified. This figure represents a typical result of the defect recognition tests.

Fewer than 20 images were used in these tests, so this study represents a very limited test. Extensive testing with hundreds or thousands of images would be necessary to determine the actual accuracy of the inspection system under production conditions. Some general conclusions can be drawn from the results, however. First, all of the products, whether defective or not, were identified with a very high

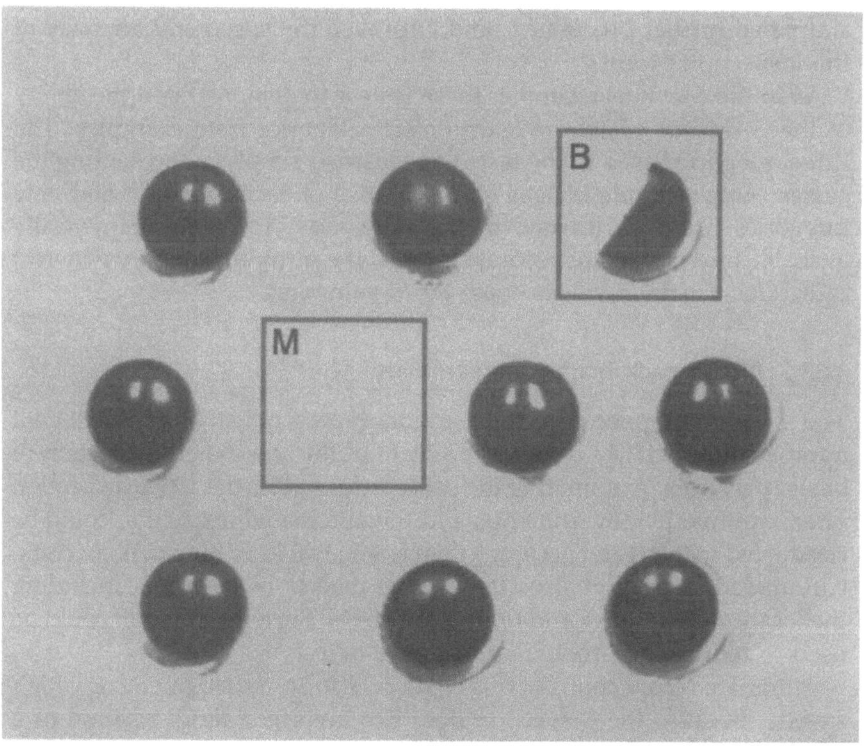

Figure 11.6 Results of blister pack inspection with broken (B) and missing (M) products identified.

degree of confidence. Because confidence levels were adjusted to avoid Type II errors (passing bad product), defects were classified with essentially 100% accuracy. Good products were also classified with a confidence level greater than 99%, so the overall accuracy of the system was excellent.

There are two primary reasons for the success of these tests. First is the fixed nature of the image acquisition process. Because the location, scale and rotation of the product were known and fixed in the images, all of the errors involved with locating, rotating and scaling the objects in the image were avoided. This led to very consistent representations being presented to the inspection system for analysis. The second attribute that led to success was the presence of excellent, controlled lighting conditions. Because of the controlled lighting used, the images of the product were very high contrast, sharp and clear. The differences between good, broken and missing products were very obvious and easy to distinguish with very high accuracy. These results simply reinforce the ideas presented in Chapter 5 about the presentation and illumination of objects for visual inspection. The presence of controlled object positioning and good lighting conditions in this example greatly simplified further processing, and improved the speed and accuracy of the inspection process.

As in the previous example, these tests also demonstrated the ability of the CAMERA system to learn defect categories from examples. The three categories used in the tests were learned simply by presenting the system with example images representative of each category, and new categories could be learned in the same way. In most commercially available blister pack inspection systems, the introduction of new defect categories usually involves some reprogramming.

11.3.3 Blister pack inspection summary

The system proposed in this section represents a more intelligent approach to the fairly common problem of the automated inspection of blister packages. Automating the procedure enables the 100% inspection often required by governments, and the automated inspection could be conducted with greater accuracy than is achievable by human inspectors. Automated inspection also allows more data to be collected, including the location, frequency and timing of defects. Such information could be used to reduce the production of future defects.

Automated inspections of this type can also be carried out at very high speeds. Because the inspection does not involve scaling, rotation or a search for the product or defects, the computationally intensive routines that perform these tasks can be eliminated from the system. The result is a very compact and reliable automated system that can perform inspections at the rate of several per second. Inspection costs would also be reduced over human inspectors, since this automated system could operate autonomously. Finally, because the automated system is once again based on artificial neural network technology, it is capable of learning new defect categories, from examples as they arise.

11.4 SUMMARY

In this chapter the practical applications of many of the techniques covered in this book have been demonstrated in both implicit and explicit inspection situations. Proper product presentation, illumination, image acquisition, and image pre-processing have been shown to greatly simplify the inspection procedure, and to improve the accuracy of the results. Edge detection and feature grouping, techniques built into the CAMERA vision model, have been used to effectively extract defect or product representations from acquired images, and those representations have been presented to an artificial neural network for recognition. The learning capability of the artificial neural network allowed the automated visual inspection system to learn new defect or product categories from visual examples. Finally, it was shown that intelligent automated visual inspection systems have many potential benefits when compared to human inspectors or commercially available inspection systems.

12

Future of automated visual inspection

12.1 INTRODUCTION

The objective of this chapter is to look into the future of automated visual inspection, and to see what impact it might have on the development of intelligent manufacturing systems. First, a system will be proposed that is capable of the flexible manufacture of printed circuit boards. The point here is not to design the best possible printed circuit board manufacturing system, but rather to use the PCB manufacturing scenario as a tool to demonstrate the variety of intelligent vision systems that would be required to support the flexible automation of even this relatively simple manufacturing process. The flexible circuit board manufacturing facility is discussed, the roles played by automated vision systems are identified, and the entire system is analyzed as an example of an intelligent system.

In the second part of this chapter a challenging automated inspection task will be introduced. This task is a real problem taken from the food products industry, and it serves to illustrate many of the shortcomings of current and proposed automated visual inspection systems. This is a challenging problem that is still beyond the capabilities of even the most advanced automated inspection systems. This type of problem provides a benchmark against which the capabilities of future inspection systems can be assessed.

12.2 PROPOSED FLEXIBLE MANUFACTURING SYSTEM

12.2.1 Flexible manufacturing

A flexible manufacturing system (FMS) typically consists of a group of automated processing stations interconnected by means of an automated material handling and storage system, and controlled by an integrated computer system. An FMS is managed, overseen and maintained by people, but it does not require direct involvement by production machine operators. Flexible manufacturing systems typically consist of a group of machine tools, robots, assembly stations and inspection stations that is capable of producing a wide variety of products.

There are many benefits to flexible manufacturing. First, a wide variety of products can be manufactured on a single group of manufacturing equipment, making short runs feasible. Short-run or single-unit products can be manufactured at rates similar to those for high quantity products. Also, the specifications or design of the product being produced can be changed on an item-by-item basis, enabling a flexible product line. Economically producing higher quality products and reducing reject costs, even with highly mixed production, is also possible with flexible manufacturing. Flexible manufacturing also enables new products, product design changes, component specification changes, and even process advancements to be quickly and economically incorporated into production. Finally, the level of computer integration of control, production and quality information commonly found in flexible manufacturing systems facilitates process monitoring, production and quality reporting, and interfaces to other systems.

12.2.2 Flexible manufacturing of printed circuit boards

The flexible manufacturing methodology must be adopted in order for manufacturing companies to take advantage of these benefits. Flexible manufacturing is applicable to a wide variety of product categories, including that of surface mount technology (SMT) printed circuit board (PCB) manufacture. Printed circuit board manufacturing is presently characterized by mass production of similar or identical products at high rates on a transfer line using predominately fixed-purpose automation. Solder paste is typically applied with the silk-screening technique, with mechanical registration of the board artwork. Tape-and-reel fed components are installed by a series of pick-and-place machines serviced by automatic feeders. Component installation is not intelligently guided, so errors and quality problems occur and are often not discoverd until final inspection. This results in high reject or rework costs. Odd components are often installed manually, resulting in variation in both the rate and quality of production as well as operator fatigue.

Changes in product design or incorporation of new components often involves extensive modification of the PCB production line, including the design and installation of new silk screens, reprogramming of pick-and-place machines, installation of new component feeders and retraining of production personnel. The introduction of a new product can often require the design and installation of a completely different production line.

In this section a flexible manufacturing system is proposed that allows the intelligent flexible production of a wide variety of printed circuit boards on an item-by-item basis if necessary. The boards may differ widely in size, component content, component density and layout. Solder paste in this proposed system is to be applied in a completely programmable manner, with a different pattern used for each board if necessary. The use of on-line automated visual inspection will minimize paste application defects. Component placement will also be completely flexible and intelligently guided, and any components not supplied by automatic tape-and-reel feeders will be identified by an intelligent vision system prior to installation to prevent incorrect component installation. Again, automated visual inspection will verify proper component orientation and positioning prior to solder reflow, minimizing rework and reject costs. Introduction of new product designs or product design changes to the system will be highly automated to minimize the engineering effort required. A possible layout for the proposed flexible PCB manufacturing cell is shown in Fig. 12.1.

In order to clarify the proposed process, the flow of a typical board through the cell will be described. First, it is assumed that unpopulated printed circuit boards, ready for component installation, will be

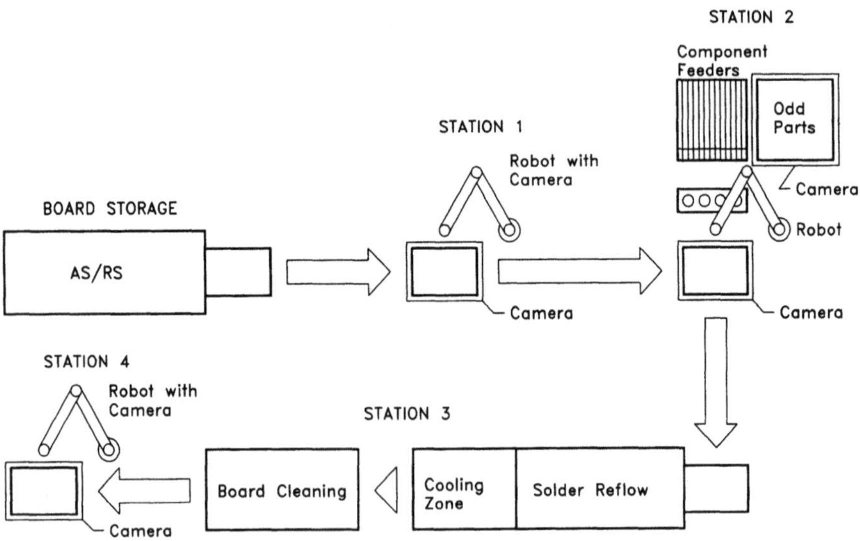

Figure 12.1 Layout of proposed flexible printed circuit board manufacturing system.

available in the automatic storage and retrieval system (ASRS). A particular board will be selected from the ASRS, based on an optimum production schedule, and the selected board will be delivered to Station 1 by the combination of an automatic conveyor and a robot manipulator. At Station 1, the board will be visually identified by a fixed overhead video camera, and the board artwork will be registered in Station 1 robot coordinates. Solder paste will then be applied to the board as necessary by the Station 1 robot, in a pre-determined optimal path for that board. The correct paste application will be verified on-line by a robot-mounted video camera. The board will then be conveyed to Station 2.

At Station 2, the board artwork will again be registered in local robot coordinates by a video camera. The surface-mount components will then be installed in the fresh solder paste by the Station 2 robot manipulator. Components will be installed in a predetermined optimum sequence for the particular board being manufactured. Robot gripper changes, if required, will be automatic. Components will be available from two areas, the feeder area and the odd part area. Components in the feeder area will be of the type typically available on tape-and-reel and will be fed by automatic component feeders. The components available from each reel will be identified by scanning a bar code on that reel. Several types of components are typically not available on tape reels, however. These components will be stored in the odd-part area on pallets or loose in bins, and they will be automatically located and identified by a fixed overhead video camera. When a component is required, it will either be available from an automatic feeder or it will be located by the vision system in the odd-part area. The component will then be picked from the proper location, oriented correctly and installed on the board. When all of the components have been installed, the board will be inspected for component location accuracy by the overhead video camera. At this stage the board will be fully assembled and ready for solder reflow at Station 3.

At Station 3, boards that have been populated and inspected as described above will be transported through an automatic vapor-phase solder reflow machine. This machine melts the solder paste, causing it to flow and make the permanent electrical connections on the board. After solder reflow the board is cooled and cleaned, and it is ready for final inspection and testing.

Final inspection and testing occurs at Station 4. Final inspection involves a fixed vision system to inspect the board for overall defects (e.g. excess solder, shorts and bridges), and a robot-mounted camera to inspect individual solder joints for quality. This visual inspection is followed by electrical testing which determines the functionality of the board. Finally, fully tested and inspected boards are packaged and shipped.

12.2.3 Visual inspection to support flexible PCB manufacturing

The manufacturing cell just described relies on several automated systems for visual inspection. At the first station, board patterns will be recognized by a visual system. Instead of having to be reprogrammed, this system will learn new board designs from examples. Board designs will be presented in the form of CAD drawings, board artwork or prototype boards. Artificial neural network technology will enable this capability. Board artwork will also be visually registered via a geometric transformation between image coordinates and robot coordinates. This will allow a robot to accurately apply solder paste to the board without any further information. A robot-mounted camera will also monitor the solder paste application process for quality.

The second station will also rely on visual identification systems. The board will again be visually registered in robot coordinates, and any odd components will be visually located and identified prior to installation on the board. The populated board will also be visually inspected to insure that the component installation process occurred correctly.

The board will also be given a final visual inspection at the last station, to determine if any obvious defects have been introduced by the soldering process, prior to functional and electrical testing. In summary, automating a flexible manufacturing process like this one will require several different kinds of intelligent visual systems, with capabilities such as object location and identification, defect identification, coordinate transformation and visual learning.

12.2.4 Expected benefits of flexible PCB manufacturing

It is difficult to estimate the economic value of the benefits that could be obtained by the development of the flexible manufacturing system described here, as it represents a change in basic philosophy in the electronics industry towards the assembly of low volume, high mix products. The expected benefits of such a system are summarized generally in the following paragraphs.

The system would reduce greatly the time to respond to product changes, which would increase the level of competitiveness. Time-based competition is becoming a reality, and manufacturing companies that can rapidly provide customers with products that meet their changing needs will prosper. This flexibility would also lead to reductions in the amount of capital invested in in-process inventories, finished goods inventories and specialized equipment.

The application of automated visual inspection systems at various points in the manufacturing process also has benefits. The costs will be reduced over manual inspection, since inspections of this variety are usually quite labor-intensive. Waste will also be reduced, since multiple automated inspections allow defective products to be removed or reworked at the most economical point in the production process. The traditional approach of inspecting products at the end of the line allows much extra value to be built into defective products.

The proposed system would greatly reduce the set-up time for assembling different products, which is composed of equipment and software set-up, due to its ability to adapt to new products. The system structure and the use of intelligent software and subsystems provides the ability to assemble even single, one-of-a-kind board, without extensive set-up procedures.

Continuous adaptation would be another important new attribute of the system. This is achieved through the use of artificial neural networks for various tasks in the system, including the support of visual inspection and recognition systems. The learning ability of artificial neural networks enables equipment and systems to adapt rapidly and economically to the demands of a continuously changing production environment.

Finally, the adoption of manufacturing practices such as those represented by this proposed system allows companies to participate in much broader markets than has been traditional. Manufacturing companies that can accommodate the production of a wide variety of products with little additional capital investment can take advantage of opportunities that may represent low production rates but high profits. In the electronics industry, for example, the production of both military and commercial products on the same line could be very advantageous.

The proposed system represents a combination of the flexibility of job shop assembly with the accuracy, quality and speed of mass production. This will both reduce the cost of the assembly and increase the level of quality that can be expected. In summary, the proposed system will reduce the cost of assembly, improve product quality and provide other benefits due to its adaptability and learning ability.

12.2.5 Flexible PCB manufacturing as an intelligent system

Consider the PCB manufacturing system proposed above against the definition of intelligent systems presented in Chapter 1. Visual, robotic and other sensors would allow the system to monitor the state of the external environment. Sensory processing would allow the system to make sense out of these inputs and to provide meaningful information for updating the internal world model. The world model would contain product information such as circuit board patterns, optimal solder paste application paths, component specifications and optimal component insertion sequences to name a few. The world model would also contain information such as production schedules, manufacturing priorities and quality goals.

The behavior generation module would generate potential action plans based on the world model and the sensory inputs. A board with a perceived quality problem, for example, could either be shipped, scrapped or reworked. The value judgement system would evaluate those alternatives and determine, based on the world model, which course of action would be most beneficial to the system. Finally, actuators would be commanded to carry out the appropriate action based on the value judgement. A board with a certain class of defect would be routed for rework, for example, rather than scrapped or shipped. In short, even this simple automated flexible manufacturing system represents all of the aspects of an intelligent system to some degree.

12.3 CHALLENGING VISUAL INSPECTION PROBLEM

The previous section described a possible future manufacturing system that incorporated many automated visual inspection systems. Although the tasks performed by these systems are difficult, they have all been accomplished in one form or another. The electronics industry has embraced automated visual inspection, and this level of interest has spurred the development and marketing of specialized inspection systems for most of the tasks described above. In contrast, this section presents an inspection problem that currently has no available off-the-shelf solution. The problem is from the baking industry, an industry where human judgement of the product is valued more than precise physical or electrical measurements. Much of the information that follows was provided by Herbert R. Tuttle, former Plant Manager at Wolferman's Bakery in Lenexa, Kansas, a division of Sara Lee Corporation.

Wolferman's manufactures 12 different varieties of English muffins at various times on a production line at the Lenexa plant. Product variations include differences in flour (wheat, whole wheat etc.), additives (cranberries, blueberries, raisins, nuts etc.), and seasonings. These varieties can differ considerably from each other in physical appearance. English muffins are produced at a rate of 120 per minute on this line. The finished products pass over a collating table where they are automatically divided into four streams which feed packaging lines. The four packaging lines each run at a rate of 30 products per minute. The current practice is to have an operator inspect the products *en masse* as they pass over the collating table, looking for and removing any defective individual products. Products removed by the inspector are manually packaged off-line and sold at 'thrift' outlets at a discount. Less than 1% of all production is currently downgraded in this manner.

It is desirable to automate this inspection for many reasons. The current inspection relies totally on subjective human judgement, and suffers from the usual problems of operator boredom and fatigue, and inconsistency over time and across product varieties. It is felt that automated inspection would result in a more consistent, higher quality product at the same or even reduced cost.

The human inspector is simultaneously judging many attributes of the product. First, the product must be the correct physical size and shape. English muffins are shaped approximately like a hockey puck with rounded corners, so the primary size attributes are height and diameter. As a practical matter, products that are too large do not fit properly in the package, and those that are too small can be disappointing to customers. Roundness of the circumference is also an issue. Products that deviate too far from being round are considered defective. The rounded shape of the corners of the product is also important. Products not shaped properly before baking can have corners that appear sharp or even pointed, and this is considered a defect.

The appearance of the surface of the muffin is also important. The product should have an overall toasted appearance, without being too dark or light. Consumers can consider products that are too light to be underdone, whereas overly dark products can be considered to be overdone or burnt. One problem with the current inspections has been that the appearance of the muffins being produced changes gradually over time, getting darker or lighter as a shift progresses. If the change is slow enough it is often not detected by the inspector. To combat this a procedure has been established where product samples are set aside every hour, and compared with the production of previous hours. The gradual shade changes become more noticeable when products made an hour or more apart are put side by side.

The surface of the product is also inspected for regularity, where dark or light spots on a muffin are considered defects. Surface texture is also important, since a product with a surface that is excessively rough or even too smooth is considered defective. English muffins are also dusted on the surface, and this dust must appear normal. Originally a residue of corn meal was used to keep the dough from sticking in the molds; the dust has become an expected part of the product.

Products with berries, nuts or raisins added undergo additional scrutiny. The additives must be dispersed properly in the product, and the correct density must be visible on the surface. Products with too many or too few visible raisins, for example, are considered defective. Some fruit additives also bleed into the surrounding dough, and excess bleeding is considered a defect.

This is a challenging inspection problem for many reasons. First, the production line is capable of producing 12 varieties, and each has its own appearance to a degree, so any inspection system would have to adapt to the product currently being produced. The most difficult problem, however, is the number of human judgements that are involved. There are currently no hard and fast rules followed for any of the defect categories, and overall appearance is considered along with the individual judgements. It is conceivable that the size and shape inspections could be automated by making visual measurements and comparing them to some model or standard. The judgement of surface color and texture is more difficult, and elaborate inspection system designs have been developed specifically for tasks like this. Most employ some form of artificial intelligence for analyzing and judging color and texture. Automatic inspection of attributes such as the distribution of raisins or the bleeding of blueberries would be even more difficult, since no standard pattern is expected, only an overall impression. Systems based on artificial neural networks have shown promise on this type of task. Because they are capable of learning from examples, they could learn to distinguish good from bad products without explicit rules, standards or models.

12.4 SUMMARY

This chapter has shown how the techniques presented in this book could be applied to the larger task of building intelligent manufacturing systems. Using printed circuit board manufacturing as an example, it has been demonstrated that automated visual systems of many varieties, and with human-like capabilities, would be required as elements of such a system. Although manufacturing systems that employ this degree of automation, intelligence and learning do not exist today, virtually all of the techniques required to make such systems a reality are either currently available or under development.

A challenging inspection problem from the food products industry was also presented. Solving this problem would require the development of an automated visual inspection system with capabilities beyond those readily available today. Being able to duplicate human qualities such as judgement, flexibility and adaptability would be necessary to fully address this problem. It is hoped that this problem represents a challenge that will encourage the additional development of intelligent visual inspection systems, and that this book will play a small part in furthering the development of systems that will some day be capable of performing inspection tasks of this level of difficulty.

Appendix A: Applications of machine vision systems for inspection in manufacturing environments

Bath soap bar inspection

A manufacturer of bar soap uses a machine vision inspection system to determine the quality of individual soap bars (Duffield, 1991). The information gathered from the vision system is sent to a personal computer, where a statistical process control software package aids in decision making.

Inspection of rivets

Rivets are inspected by a machine vision system at the rate of 600 per minute (Hollingum, 1984). Two cameras are used to inspect the stem of the rivet, and one to inspect the head. Fiber-optic light sources are used to provide uniform illumination. Rivets were previously manually inspected by sampling batches, where an entire batch was discarded if it was contaminated with bad product. The vision-based system allows 100% inspection, and rejection of individual rivets.

Automatic grading of hardwood lumber

A system for the automatic grading of rough hardwood lumber has been developed (Conners *et al.*, 1992). The system images rough hardwood boards as they are moved past a camera, and features representing defects are extracted. The size and frequency of the defects determines the grade of the board. It is hoped that this system will evolve into a general-purpose tool for the forest products industry.

Inspection of photographic film

Machine vision is being used for the 100% inspection of photographic film for surface imperfections (Hollingum, 1984). Since the film is obviously light sensitive, the process uses an infra-red laser to scan the surface of the film. Visual data is then analyzed to locate surface defects. Defect information is then used to guide the web cutting process, so only good product is shipped and defective areas are rejected.

Automatic grading of wood slabs

A machine vision system has been developed for the automatic grading of beechwood slabs prior to their manufacture into parquet flooring (Ersboll and Conradsen, 1992). This task is almost exclusively done by human inspectors in the industry, and human inspectors are difficult to obtain and require a long training period. The fact that wood is a natural material with great variations makes this task particularly difficult. The vision system is used to extract features, and the combination of these features found on a particular panel is used to grade the panels into one of five categories. A prototype of this system has been built and is being tested in a factory.

Verification of bolt size

The proper size and thread pitch of bolts in automotive cylinder block assembly manufacture is being verified by a machine vision system (Hollingum, 1984). The assembly process for installing bearing caps in engine blocks involves the insertion of bolts, followed by tightening. Before these two stages, a machine vision system verifies the proper size and thread pitch of the bolt being installed. Installation of improper bolts can result in very expensive damage to the engine block assembly.

Surface mount device placement inspection

A system has been developed for the automatic measurement of surface mount device placement on printed circuit boards (Capson and Tsang, 1990). Infra-red illumination is used to image component leads and solder pads, and determine their relative positions. Displacement and angular orientation of the component relative to the pads is then computed. The results compare favorably with stated placement accuracies and with manual measurements.

Detection and classification of surface defects in metal

A vision system was employed for 100% inspection, in real time, of metal being rolled (Neubauer, 1991). A grey-scale image of the surface was analyzed using texture segmentation, and the results of the analysis were fed into a neural network for identification. The average success rate for proper identification and classification of a defect was 95%.

Inspection and gauging of car bodies

A system has been applied to the three-dimensional measurement of automobile bodies (Ye *et al.*, 1992). The system uses structured light, where a pattern of light is projected onto the car and a grey-scale camera is used to visually measure the location of critical points. The system has proved successful in the accurate non-contact measurement of free-form surfaces.

On-line measurement of steel bar

A vision system is being used to measure the dimensions of hot steel bar during manufacture (Hollingum, 1984). The bar is back-lit and both the light source and the camera rotate in order to inspect the complete perimeter of the bar. The measurement is taken approximately every 18 degrees of rotation, so the diameter of round bars, as well as flat-to-flat and diagonal measurements of other shapes, can be repeatedly verified. The equipment is water cooled so it can operate in the high-temperature environment, and it consistently outperforms an earlier measurement technique.

Automated measurement of drill wear

Automated measurements of flank wear on drills used for printed circuit board drilling were accomplished via machine vision (Ramirez and Thornhill, 1992). Flank area wear was visually measured rather than using the more common flank width measurement, and this method proved to be more effective. The vision-based system was faster and more accurate than previous methods.

Aircraft skin scratch measurement

Machine vision provides a viable alternative to manual inspection of aircraft skin material (Sarr, 1991). Aircraft skin must be free of structural defects and surface imperfections, and individual defects must be accurately measured and recorded. A vision system which analyzes microscopic structures from laser images is used, and hardware and software has been developed which enables the system to measure the depth of a scratch. The system has proved to be highly repeatable, and it is suitable for use in an aircraft manufacturing environment or on existing planes.

Determination of lighter flint-wheel orientation

A vision system has been used to determine the orientation of flint wheels prior to their assembly into cigarette lighters (Safabakhsh, 1991). The assembly process requires the assembly of the flint wheel in a certain orientation. Computer vision enables the assembly system to determine proper orientation for both upright and fallen wheels and properly insert them. Features are extracted from wheel images which provide the required distinguishability. The system has proved to be very accurate, and it operates in near-real time.

Appendix B

Proof of convergence of the learning algorithm employed in the HAVNET artificial neural network

Specifically, it is necessary to prove

$$\lim_{t \to \infty} \left[\Delta w\right] = 0 \quad \text{and} \quad \lim_{t \to \infty} \left[w\right] = k \qquad \forall w \;\middle|\; w \in W \quad \text{(B.1)}$$

where t is the number of training passes, W is the set of weights for a given node, and k is a constant.

PROOF

Given a finite set of input patterns A^m, $m = 1 \ldots M$, to be classified by a node, let the set G be the set of all weights for that node such that the corresponding input is 1 in at least one input pattern:

$$G = \left\{ w^n_{(x+\delta),\,(y+\delta)} \;\middle|\; \exists a^m_{x,y}\; a^m_{x,y} = 1 \right\} \qquad m = 1 \ldots M \text{ (B.2)}$$

Let the set H be the set of all weights for the node such that the corresponding input is 0 in all input patterns:

$$H = \left\{ w^n_{(x+\delta),\,(y+\delta)} \;\middle|\; \forall a^m_{x,y}\; a^m_{x,y} = 0 \right\} \qquad m = 1 \ldots M \text{ (B.3)}$$

It follows then that these two sets G and H are mutually exclusive:

$$G \cap H = \varnothing \qquad \text{(B.4)}$$

It also follows that each member of the weight set W must be an element of either the set G or the set H:

$$\forall w^n_{(x+\delta),(y+\delta)} \;\Big|\; w^n_{(x+\delta),(y+\delta)} \in W$$

$$w^n_{(x+\delta),(y+\delta)} \in G \vee w^n_{(x+\delta),(y+\delta)} \in H \tag{B.5}$$

Furthermore, the union of the sets G and H equals the set W:

$$W = G \cup H \tag{B.6}$$

The proof consists of four cases. Each case assumes that all weights are initialized at zero. Superscripts representing the node number, and subscripts representing position in the matrix, are dropped from the weights at this point for reasons of clarity.

Case 1

By the definition of the learning rule, any weight w that is a member of the set H undergoes no change:

$$\forall t \forall w \in H \; \Delta w = 0 \tag{B.7}$$

Therefore convergence is assured for this case:

$$\lim_{t\to\infty} \left[\Delta w\right] = 0 \qquad \lim_{t\to\infty} \left[w\right] = 0 \qquad \forall w \;\Big|\; w \in H \tag{B.8}$$

Case 2

In the special case where $\alpha = 0$, any weight w that is a member of G undergoes no change:

$$\Delta w = \alpha(1 - 0) = 0 \tag{B.9}$$

Therefore convergence is assured for this case:

$$\lim_{t\to\infty} \left[\Delta w\right] = 0 \qquad \lim_{t\to\infty} \left[w\right] = 0 \qquad \forall w \;\Big|\; w \in G \tag{B.10}$$

Case 3

In the special case where $\alpha = 1$, any weight w that is a member of the set G becomes 1 after a single training pass:

$$\Delta w = \alpha(1 - 0) = 1 \qquad w(1) = 1 + 0 = 1 \tag{B.11}$$

Furthermore, any weight w undergoes no further changes after the first training pass:

$$\Delta w = \alpha(1 - w) = \alpha(1 - 1) = 0 \tag{B.12}$$

Therefore convergence is assured for this case:

$$\lim_{t\to\infty} \left[\Delta w\right] = 0 \qquad \lim_{t\to\infty} \left[w\right] = 1 \qquad \forall w \mid w \in G \tag{B.13}$$

Case 4

In the general case where $0 < \alpha < 1$, any weight w that is a member of the set G will undergo the following minimum change during each training pass:

$$\Delta w = \alpha(1 - w) \tag{B.14}$$

The minimum weight changes generated by the first four training passes are shown below:

$$\begin{aligned}
t &= 1: \quad \Delta w = \alpha \\
t &= 2: \quad \Delta w = \alpha - \alpha^2 \\
t &= 3: \quad \Delta w = \alpha - 2\alpha^2 + \alpha^3 \\
t &= 4: \quad \Delta w = \alpha - 3\alpha^2 + 3\alpha^3 - \alpha^4
\end{aligned}$$

The terms of the binomial expansion can be extracted from the above equations, and a general form for the minimum weight change at each training iteration can be developed as follows:

$$\Delta w(t) = \alpha(1 - \alpha)^{t-1} \tag{B.15}$$

As training proceeds,

$$\lim_{t\to\infty} \left[\Delta w\right] = \lim_{t\to\infty} \left[\alpha(1 - \alpha)^{t-1}\right] = \alpha \lim_{t\to\infty} \left[(1 - \alpha)^{t-1}\right] = 0 \tag{B.16}$$

since $0 < \alpha < 1$.

The weight change is guaranteed to converge to zero. Since the weight changes are additive in nature, the expression for the minimum value of any weight w after t training iterations is

$$w(t) = \sum_{i=1}^{t} \alpha(1 - \alpha)^{i-1} \tag{B.17}$$

This is the familiar infinite geometric series, and it is known to obey the following convergence formula under these conditions:

$$\lim_{t\to\infty} \left[w(t)\right] = \frac{\alpha}{1 - (1 - \alpha)} = \frac{\alpha}{\alpha} = 1 \tag{B.18}$$

Therefore convergence is assured for this case:

$$\lim_{t\to\infty}\left[\Delta w\right] = 0 \qquad \lim_{t\to\infty}\left[w\right] = 1 \qquad \forall w \;\Big|\; w \in G \tag{B.19}$$

Since the sets G and H make up the set W (equation B.6), convergence is assured for all cases:

$$\lim_{t\to\infty}\left[\Delta w\right] = 0 \qquad \lim_{t\to\infty}\left[w\right] = k \qquad \forall w \;\Big|\; w \in W \tag{B.20}$$

where $k \in \{0, 1\}$.

End of proof.

References

Abramov, I. (1968) Further analysis of the response of LGN cells. *Journal of the Optical Society of America*, **58**, 574–579.

Adaptive Solutions (1995) *CNAPS/PC: Parallel Co-Processor Board for the PC*, product information sheet, Adaptive Solutions, Beaverton, OR.

Albus, J.S. (1991) Outline for a theory of intelligence. *IEEE Transactions on Systems, Man, and Cybernetics*, **21**(3), 473–509.

Allen, P.K. (1987) *Robotic Object Recognition Using Vision and Touch*, Kluwer, Boston, MA.

Attneave, F. (1950) Dimensions of similarity. *American Journal of Psychology*, **63**, 516–556.

Attneave, F. (1954) Some informational aspects of visual perception. *Psychological Review*, **61**, 183–193.

Attneave, F. (1957) Physical determinants of the judged complexity of shapes. *Journal of Experimental Psychology*, **53**, 221–227.

Attneave, F. and Arnoult, M.D. (1966) The quantitative study of shape and pattern perception, in *Pattern Recognition* (ed. L. Uhr), Wiley, Chichester.

Bains, N. and David, F. (1990) Machine vision inspection of fluorescent lamps, in *Proceedings of the Conference on Machine Vision Systems Integration in Industry*, SPIE Vol. 1386, pp. 232–242.

Barrow, H. and Tenenbaum, J. (1981) Interpreting line drawings as three-dimensional surfaces. *Artificial Intelligence*, **17**, 75–116.

Batchelor, B.G. and Braggins, D.W. (1992) Commercial vision systems, in *Computer Vision: Theory and Industrial Applications* (ed. Torras), Springer-Verlag, New York, pp. 405–452.

Bergevin, R. and Levine, M.D. (1992a) Extraction of line drawing features for object recognition. *Pattern Recognition*, **25**, 319–334.

Bergevin, R. and Levine, M.D. (1992b) Part decomposition of objects from single view line drawings. *Computer Vision, Graphics, and Image Processing: Image Understanding*, **55**, 73–83.

Bergevin, R. and Levine, M.D. (1993) Generic object recognition: Building and matching coarse descriptions from line drawings. *IEEE Transactions on Pattern Analysis and Machine Intelligence*, **15**(1), 19–36.

Biederman, I. (1985) Human image understanding: Recent research and a theory. *Computer Vision, Graphics, and Image Processing*, **32**, 29–73

Brooks, M.J. and Horn, B.K.P. (1985), Shape and source from shading, in *Proceedings of the International Joint Conference on Artificial Intelligence*, Los Angeles, Aug. 18–23. pp. 932–936.

Brooks, R.A. (1980) Goal-directed edge linking and ribbon finding, in *Proceedings of the ARPA Image Understanding Workshop*, pp. 95–103.

Brooks, R.A. (1981) Symbolic reasoning among 3-D models and 2-D images. *Artificial Intelligence*, **17**, 285–348.

Bruce, V. and Green, P. (1985) *Visual Perception: Physiology, Psychology and Ecology*, Lawrence Erlbaum Associates.

Bundsen, C. and Larsen, A. (1975) Visual transformation of size. *Journal of Experimental Psychology: Human Perception and Performance*, **1**, 214–220.

Burr, D.J. (1987) Experiments with a connectionist text reader, in *Proceedings of the IEEE First International Conference on Neural Networks*, Vol. 4, pp. 717–724.

Capson, D.W. and Tsang, R.M.C. (1990) An experimental system for SMD component placement inspection, in *Proceedings of the IEEE International Conference on Industrial Electronics, Control and Instrumentation (IECON) '90* IEEE Vol. 1, pp. 815–820.

Carpenter, G.A. and Grossberg, S. (1987a) A massively parallel architecture for a self-organizing neural pattern recognition machine. *Computer Vision, Graphics and Image Processing*, **37**, 54–115.

Carpenter, G.A. and Grossberg, S. (1987b) ART-2: Self-organization of stable category recognition codes for analog input patterns. *Applied Optics*, **26**(23), 4919–4930.

Carpenter, G.A. and Grossberg, S. (1988) The ART of adaptive pattern recognition by a self-organizing neural network. *Computer*, **21**, 77–88.

Carpenter, G.A., Grossberg, S., Markuzon, N., *et al.* (1992) Fuzzy ARTMAP: A neural network architecture for incremental supervised learning of analog multidimensional maps. *IEEE Transactions on Neural Networks*, **3**(5), 698–713.

Carpenter, G.A., Grossberg, S. and Mehanian, C. (1989) Invariant recognition of cluttered scenes by a self-organizing ART architecture: CORT-X boundary segmentation. *Neural Networks*, **2**(3), 169–181.

Carpenter, G.A., Grossberg, S. and Reynolds, J.H. (1991) ARTMAP: Supervised real-time learning and classification of nonstationary data by a self-organizing neural network. *Neural Networks*, **4**, 565–588.

Carpenter, G.A., Grossberg, S. and Rosen, D.B. (1991a) ART-2A: An adaptive resonance algorithm for rapid category learning and recognition. *Neural Networks*, **4**, 493–504.

Carpenter, G.A., Grossberg, S. and Rosen, D.B. (1991b) Fuzzy ART: Fast stable learning and categorization of analog patterns by an adaptive resonance system. *Neural Networks*, **4**, 759–771.

Casperson, R.C. (1950) The visual discrimination of geometric forms. *Journal of Experimental Psychology*, **40**, 668–681.

Chan, J.P., Batchelor, B.G., Harris, I.P. and Perry, S.J. (1990) Intelligent visual inspection of food products, in *Proceedings of the Conference on Machine Vision Systems Integration in Industry*, SPIE Vol. 1386, pp. 171–179.

Conners, R.W. *et al.* (1992) Machine vision system for automatically grading hardwood lumber. *Industrial Metrology*, **2**(3–4), 317–342.

Cottrell, G.W., Munro, P. and Zisper, D. (1987) *Image Compression by Backpropagation: An Example of Extensible Programming*, Technical Report ICS 8702, Institute of Cognitive Science, University of California, San Diego.

Csaszar, A. (1978) *General Topology*, Adam Hilger.

D'Agostino, S. (1988) Produce sizing and grading using machine vision, in *Proceedings of SME Robots 12 and Vision '88*, Detroit, June 1988, Vol. 11, pp. 1–14.

Dagli, C.H. and Rosandich, R.G. (1993) A compact neocognitron for the location and recognition of electronic components in captured video images, in

Proceedings of the World Congress on Neural Networks, Portland, OR, July 11–15. Lawrence Erlbaum Associates, Hillside, NJ, Vol. III, pp. 700–703.

Dagli, C.H. and Vellanki, M. (1992) Automated precision assembly through neuro-vision, in *Proceedings of Applications of Neural Networks III*, SPIE Vol. 1709, pp. 493–504.

Dagli, C.H., Fernandez, B.R., Ghosh, J. and Kumara, R.T.S. (eds) (1994) *Intelligent Engineering Systems Through Artificial Neural Networks Vol. 4*, ASME Press, New York.

Dagli, C.H., Akay, M., Chen, C.L. *et al.* (eds) (1995) *Intelligent Engineering Systems Through Artificial Neural Networks Vol. 5*, ASME Press, New York.

Dahle, O., Sommerfelt, A. and McLeod, A. (1990) On-line inspection of extruded profile geometry, in *Proceedings of SME Vision '90*, Detroit, Nov. 1990, Vol. 6, pp. 1–9.

Dearborn, G.V.N. (1899) Recognition under objective reversal. *Psychology Review*, **6**, 395–406.

Dev, P. (1975) Perception of depth surfaces in random-dot stereograms: A neural model. *International Journal of Man-Machine Studies*, **7**, 511–528.

De Valois, R.L., Smith, C.J., Kitai, S.T. and Kardy, A.J. (1958) Response of single cells in monkey lateral geniculate nucleus to monochromatic light. *Science*, **127**(3292), 238–239.

Dodwell, P.C. (ed) (1971) *Perceptual Processing: Stimulus Equivalence and Pattern Recognition*, Appleton-Century-Crofts.

Dorf, R.C. (ed) (1988) *International Encyclopedia of Robotics: Applications and Automation*, Wiley, New York.

Dowling, J.E. (1968) Synaptic organization of the frog retina: An electron microscope analysis comparing the retinas of frogs and primates, *Proceedings of the Royal Society of London B*, **170**, 205–228.

Dreyfus, D.D. (1989) Comments in panel discussion: Is Industry Ready for Machine Vision, in *Machine Vision for Inspection and Measurement* (ed. H. Freeman), Academic Press, New York, pp. 223–236.

Duda, R. and Hart, P. (1973) *Pattern Classification and Scene Analysis*, Wiley, New York.

Duetsch, J.A. (1955) A theory of shape recognition. *British Journal of Psychology*, **46**, 30–37.

Duffield, S. (1991) Barring bad bars with SQC. *Manufacturing Systems*, **9**(10), 44–45.

Enroth-Cugell, C. and Robson, J.G. (1966) The contrast sensitivity of retinal ganglion cells of the cat. *Journal of Physiology*, **187**, 517–552.

Ersboll, B.K. and Conradsen, K. (1992) Automated grading of wood slabs. *Industrial Metrology*, **2**(3–4), 219–236.

Frankot, R.T. and Chellappa, R. (1990) Estimating surface topography from SAR imagery using shape from shading techniques. *Artificial Intelligence*, **43**(3), 271.

Freeman, J.A. and Skapura, D.M. (1991) *Neural Networks: Algorithms, Applications, and Programming Techniques*, Addison-Wesley, Reading, MA.

Fukushima, K. (1980) Neocognitron: A self-organizing neural network model for a mechanism of pattern recognition unaffected by shift in position. *Biological Cybernetics* (April), 193–202.

Fukushima, K. (1988) A neural network for visual pattern recognition. *Computer*, pp. 65–75.

Fukushima, K., Miyake, S. and Ito, T. (1983) Neocognitron: A neural network model for a mechanism of pattern recognition. *IEEE Transactions on Systems, Man and Cybernetics*, **13**(5), 826–834.

Galindo, P.L. and Michaux, T. (1990) An improved competitive learning algorithm applied to high level speech processing, in *Proceedings of the International Joint Conference on Neural Networks*, Vol. 1, pp. 142–146.

Gibson, E.J., Gibson, J.J., Smith, O.W. and Flock, H.R. (1959) Motion parallax as a determinant of perceived depth. *Journal of Experimental Psychology*, **58**, 40–51.

Gibson, J.J. (1950) *Perception of the Visual World*, Houghton Mifflin, New York.

Gibson, J.J. (1966) *The Senses Considered as Perceptual Systems*, Houghton Mifflin, Boston.

Gjerdingen, R.O. (1990) Categorization of musical patterns by self-organizing neuron-like networks. *Music Perception*, **7**, 339–370.

Glover, D.E. (1988) Optical fourier/electronic neurocomputer machine vision inspection system, in *Proceedings of the SME Vision '88 Conference*, Detroit, MI, June 1988.

Grossberg, S. (1983) The quantized geometry of visual space: The coherent computation of depth, form, and lightness. *Behavioral and Brain Sciences*, **6**, 625–692.

Grossberg, S. and Mingolla, E. (1985) Neural dynamics of form perception: Boundary completion, illusory figures, and neon color spreading. *Psychological Review*, **92**, 173–211.

Grossberg, S. and Mingolla, E. (1987) Neural dynamics of surface perception: Boundary webs, illuminants, and shape-from-shading. *Computer Vision, Graphics, and Image Processing*, **37**, 116–165.

Grossberg, S., Mingolla, E. and Tordorovic, D. (1989) A neural network architecture for preattentive vision. *IEEE Transactions on Biomedical Engineering*, **36**(1), 65–84.

Guyon, I., Vapnik, V., Boser, B. *et al.* (1992) Structural risk minimization for character recognition, in *Advances in Neural Information Processing, Vol. 4* (eds J. Moody, S. Hanson and R. Lippmann), Morgan Kauffman, San Mateo, CA, pp. 471–479.

Haber, R.N. (1983) Stimulus information and processing mechanisms in visual space perception, in *Human and Machine Vision* (eds J. Beck, B. Hope and A. Rosenfeld), Academic Press, New York.

Hake, H.W. (1966) Form discrimination and the invariance of form, in *Pattern Recognition* (ed L. Uhr), Wiley, Chichester.

Halliday, D. and Resnick, R. (1974) *Fundamentals of Physics*, Wiley, Chichester.

Harai, Y. and Fukushima, K. (1978) An inference upon the neural network finding binocular correspondence. *Biological Cybernetics*, **31**, 209–217.

Harms, T.M. (1992) Machine vision: What can it do for you, in *Proceedings of the Annual Conference of Electrical Engineering Problems in the Rubber and Plastics Industries*, IEEE Publications, pp. 30–34.

Harnarine, R. and Mahabir, M. (1990) A database system for parts verification and defects reporting on assembly lines, in *Proceedings of Manufacturing International '90*, Atlanta, GA, March 1990. Vol. V, pp. 75–81.

Harvey, R.L., DiCaprio, P.N., Heinemann, K.G. *et al.* (1990) A neural architecture for visual recognition of general stationary objects by machines, in *Proceedings of the International Joint Conference on Neural Networks*, Vol. 2, pp. 937–942.

Hebb, D.O. (1949) *The Organization of Behavior*, Wiley, New York.

Hedengren, K. (1986) Automatic image analysis for inspection of complex objects: A systems approach, in *Proceedings of the Eighth International Conference on Pattern Recognition*, Paris, pp. 919–921.

Helson, H. (1938) Fundamental principles in color vision: The principle governing changes in hue, saturation, and lightness of non-selective samples in chromatic illumination. *Journal of Experimental Psychology*, **23**, 439–471.

Hillis, W.D. (1985) *The Connection Machine*, MIT Press, Cambridge, MA.

Hochberg, J. and Brooks, V. (1960) The psychophysics of form: Reversible perspective drawings of spatial objects. *American Journal of Psychology*, **73**, 337–354.

Hochberg, J. and Brooks, V. (1962) Pictorial recognition as an unlearned ability: A study of one child's performance. *American Journal of Psychology*, **75**, 624–628.

Hochberg, J. and McAlister, E. (1953) A quantitative approach to figural goodness. *Journal of Experimental Psychology*, **46**, 361–364.

Hodges, A. (1983) *Alan Turing: the Enigma*, Simon & Schuster, New York.

Hoffman, D.D. and Richards, W. (1985) Parts of recognition. *Cognition*, **18**, 65–96.

Hollingum, J. (1984) *Machine Vision: The Eyes of Automation*, IFS Publications, Bedford, UK.

Hopfield, J.J. and Tank, D.W. (1985) Neural computation of decisions in optimization problems. *Biological Cybernetics*, **52**, 141–152.

Horn, B.K.P. (1977) Image intensity understanding. *Artificial Intelligence*, **8**(2), 201–231.

Horn, B.K.P. (1986) *Robot Vision*, MIT Press, Cambridge, MA.

Hubel, D.H. and Wiesel, T.N. (1962) Receptive fields, binocular interaction and functional architecture in the cat's visual cortex. *Journal of Physiology*, **160**, 106–154.

Hubel, D.H. and Wiesel, T.N. (1977) Functional architecture of macaque monkey visual cortex. *Proceedings of the Royal Society of London B*, **198**, 1–59.

Huttenlocher, D.P., Kedem, K. and Sharir, M. (1990) The upper envelope of Voronoi surfaces and its applications, in *Proceedings of the Sixth ACM Conference on Computational Geometry*, pp. 194–293.

Huttenlocher, D.P., Klanderman, G.A. and Rucklidge, W.J. (1993) Comparing images using the Hausdorff distance. *IEEE Transactions on Pattern Analysis and Machine Intelligence*, **15**(9), 850–863.

Ikeuchi, K. and Horn, B.K.P. (1981) Numerical shape from shading and occluding boundaries. *Artificial Intelligence*, **17**(1–3), 141–184.

Ittleson, W.H. (1952) *The Ames Demonstrations in Perception*, Hafner, New York.

Judd, D.B. (1940) Hue saturation and lightness of surface colors with chromatic illumination. *Journal of the Optical Society of America*, **30**, 2–32.

Judd, D.B. (1960) Appraisal of Land's work on two-primary color projections. *Journal of the Optical Society of America*, **50**, 254–268.

Julesz, B. (1971) *Foundations of Cyclopean Perception*, University of Chicago Press, Chicago.

Julesz, B. and Chang, J.J. (1976) Interaction between pools of binocular disparity detectors tuned to different disparities. *Biological Cybernetics*, **22**, 107–120.

Kassler, M., Corke, P.I. and Wong, P.C. (1991) Automatic grading and packing of prawns, in *Proceedings of the International Robots and Vision Automation Conference*, Detroit. Vol. 9, pp. 27–32.

Kennedy, C.W., Hoffman, E.G. and Bond, S.D. (1987) *Inspection and Gauging*, 6th edn, Industrial Press, New York.

Kishi, T., Hibara, T. and Nakano, N. (1993) CPT inspection systems with image processing, in *Proceedings of the IEEE International Conference on Industrial Electronics, Control, and Instrumentation (IECON '93)*, Maui, HA, Nov. 1993. pp. 1893–1897.

Koenderink, J.J. and Van Dorn, A.J. (1976) Geometry of binocular vision and a model for stereopsis. *Biological Cybernetics*, **29**, 29–35.
Koenderink, J.J. and Van Doorn, A.J. (1979) The internal representation of solid shape with respect to vision. *Biological Cybernetics*, **32**, 211–216.
Koffka, K. (1935) *Principles of Gestalt Psychology*, Harcourt-Brace.
Kohonen, T. (1984) *Self-Organization and Associative Memory*, Springer-Verlag, New York.
Kohonen, T. (1988) The Neural Phonetic Typewriter, *Computer*, **21**(3), 11–22.
Konig, P. and Schillen, T.B. (1991) Stimulus-dependent assembly formation of oscillatory responses: I. Synchronization, *Neural Computation*, **3**, 155.
Krauskopf, J., Duryea, RA. and Bitterman, M.E. (1954) Threshold for visual form: further experiments, *American Journal of Psychology*, **67**, 427–440.
Kuffler, S.W. (1953) Discharge patterns and functional organization of mammalian retina, *Journal of Neurophysiology*, **16**(1), 37–68.
Kurowski, C.W. (1988) Machine vision at Volkswagen Westmoreland, in *Proceedings of SME Robots 12 and Vision '88*, Detroit, MI, June 1988. Vol. 455, pp. 1–21.
Laligant, O., Truchetet, F. and Fauvet, E. (1993) Wavelets transform in artificial vision inspection of threading, in *Proceedings of the IEEE International Conference on Industrial Electronics, Control, and Instrumentation (IECON '93)*, Maui, HA, Nov. 1993. pp, 513–518.
Land, E.H. and McCann, J.J. (1971), Lightness and retinex theory, *Journal of the Optical Society of America*, **61**, 1–11.
Lee, K.J., Lim, H.K., Hong, C.K. and Yang, W.S. (1991) A vision system for video head inspection, in *Proceedings of the IEEE International Conference on Industrial Electronics, Control and Instrumentation (IECON '91)*. IEEE pp. 986–991.
Lee, M.H. (1989) *Intelligent Robotics*, Open University Press, Milton Keynes, UK.
Lehar, S. (1993a) Orientational harmonic model for illusory boundary formation in biological vision, in *Proceedings of the World Congress on Neural Networks*, Portland, OR, July 11–15. Lawrence Erlbaum Associates, Hillside, NJ, Vol. I, pp. 76–79.
Lehar, S. (1993b) Orientational harmonic model for rotation, translation, and scale invarient pattern recognition in biological vision, in *Proceedings of the World Congress on Neural Networks*, Portland, OR, July 11–15. Lawrence Erlbaum Associates, Hillside, NJ, pp. 80–83.
Levine, M.D. (1985) *Vision in Man and Machine*. McGraw-Hill, New York.
Lindsay, P.H. and Norman, D.A. (1972) *Human Information Processing, an Introduction to Psychology*. Academic Press, New York.
Lowe, D.G. (1985) *Perceptual Organization and Visual Recognition*. Kluwer Academic Publishers, Hingham, MA.
Mach, E. (1914) *The Analysis of Sensations*. Open Court, Chicago.
Mair, G.M. (1988) *Industrial Robotics*, Prentice-Hall, New York.
Malik, J. and Maydan, D. (1989) Recovering three-dimensional shape from a single image of curved objects, *IEEE Transactions on Pattern Analysis and Machine Intelligence*, **11**(6), 555.
Marks, W.B., Dobelle, W.H. and MacNichol, E.F. (1964) Visual pigments of single primate cones, *Science*, **143**, 1181–1183.
Marr, D.C. (1977) Analysis of occluding contour, *Proceedings of the Royal Society of London B*, **197**, 441–475.
Marr, D.C. (1979) Representing and computing visual information, in *Artificial Intelligence: An MIT Perspective* (eds P.H. Winston and R.H. Brown). MIT Press, Cambridge, MA.

Marr, D.C. (1980) Visual information processing: the structure and creation of visual representations. *Philosophical Transactions of the Royal Society of London B*, **290**, 199–218.
Marr, D.C. (1982) *Vision*, Freeman Publishing, New York.
Marr, D.C. and Hildreth, E. (1980) Theory of edge detection, *Proceedings of the Royal Society of London B*, **207**, 187–217.
Marr, D.C. and Nishihara, H.K. (1977) Representation and recognition of the spatial organization of three-dimensional shapes, *Proceedings of the Royal Society of London B*, **200**, 269–294.
Marr, D.C. and Poggio, T. (1976) Cooperative computation of stereo disparity, *Science*, **194**, 283–287.
Marshall, A.D. (1990) Automatically inspecting gross features of machined objects using three-dimensional depth data, in *Proceedings of the Conference on Machine Vision Systems Integration in Industry*, SPIE Vol. 1386, pp. 243–254.
Marshall, A.D. and Martin, R.R. (eds) (1992), *Computer Vision, Models, and Inspection*, World Scientific Publishing, Singapore.
McCulloch, W.S. and Pitts, W.H. (1945) A logical calculus of the ideas immanent in nervous activity, *Bulletin of Mathematical Biophysics*, **7**, 89–93.
Melville, R.C. (1985) An implementation study of two algorithms for the minimum spanning circle problem, in *Computational Geometry* (ed. G.T. Toussaint). Elsevier Amsterdam.
Miller, S. (1985) Recent developments in robotics and flexible manufacturing systems, in *Robotics and Flexible Manufacturing Technologies* (eds R.U. Ayres, *et al.*), Noyes Publications, Park Ridge, NJ.
Milton-Bradley Co. (1988) *Puzzle #4264: Wooden Map of the United States*, Milton-Bradley, Springfield, MA.
Minsky, M. and Papert, S. (1969) *Perceptrons*. MIT Press, Cambridge, MA.
Moir, P.W. (1989) *Efficient Beyond Imaging: CIM and Its Applications for Today's Industry*, Ellis Horwood, Chichester, UK.
Mundy, J., Noble, A., Marinos, C. *et al.* (1992) An object-oriented approach to template guided visual inspection, in *Proceedings of the IEEE 1992 Conference on Computer Vision and Pattern Recognition*, Champaign, IL, June 1992. pp. 386–392.
Narasimhan, M.A. (1992) Computer vision for electronics manufacturing, in *Proceedings of the IEEE International Conference on Industrial Electronics, Control, and Instrumentation (IECON '92)*, San Diego, CA, Nov 1992. IEEE pp. 1030–1040.
Nelson, J.I. (1975), Globality and stereoscopic fusion in binocular vision, *Journal of Theoretical Biology*, **49**, 1–88.
Neubauer, C. (1991) Fast detection and classification of defects on treated metal surfaces using a backpropagation neural network, in *Proceedings of the International Joint Conference on Neural Networks*, pp. 1148–1153.
Nevatia, R. and Babu, K.R. (1980) Linear feature extraction and description, *Computer Graphics and Image Processing*, **13**, 257–269.
Newman, T.S. (1993) *Experiments in 3D CAD-based inspection using range images*, Ph.D. thesis, Department of Computer Science, Michigan State University.
Newman, T.S. and Jain, A.K. (1994) CAD-based inspection of 3D objects using range images, in *Proceedings of Second IEEE CAD-Based Vision Workshop*, Champion, PA, Feb 1994. pp. 236–243.
Newman, T.S. and Jain, A.K. (1995) A survey of automated visual inspection. *Computer Vision and Image Understanding*, **61**(2), 231–262.
Noble, A. Nguyen, V.D., Martinos, C. *et al.* (1992) Template guided visual

inspection, in *Proceedings of the Second European Conference on Computer Vision*, Santa Margherita Ligure, Italy, May 1992. pp. 893–901.

Novini, A.R. (1990) Fundamentals of machine vision inspection in metal container and glass manufacturing, in *Proceedings of SME Vision '90*, Detroit, MI, Nov 1990. Vol. 14, pp. 1–11.

Olympieff, S., Pineda, J.C. and Horaud, P. (1982) The use of vision for industrial inspection, in *Proceedings of the Sixth International Conference on Automated Inspection and Product Control*, Birmingham, UK. pp. 179–190.

Palmer, S.E. (1983) The psychology of perceptual organization: A transformational approach, in *Human and Machine Vision* (eds J. Beck, B. Hope and A. Rosenfeld), Academic Press, New York.

Park, J.-S. and Tou, J.T. (1990) A solder joint inspection system for automated printed circuit board manufacturing, in *Proceedings of the IEEE International Conference on Robotics and Automation*, Cincinnati, OH. pp. 1290–1295.

Parker, D.B. (1982) *Learning Logic.* Invention Report 581–64, File 1, Office of Technology Licensing, Stanford University.

Pearson, D.E., Rubinstein, C.B. and Spivack, G.J. (1969) Comparison of perceived color in two-primary computer generated artificial images with predictions based on the Helson-Judd formulation, *Journal of the Optical Society of America*, **59**, 644–658.

Pitts, W.H and McCulloch, W.S. (1947) How we know universals: the perception of auditory and visual forms, *Bulletin of Mathematical Biophysics*, **9**, 127–147.

Rak, S. and Kolodzy, P. (1988) Invariant object recognition with the Adaptive Resonance (ART) network, *Neural Networks Supplement: INNS Abstracts*, **1**, 461.

Ramirez, C.N. and Thornhill, J.R. (1992) Automated measurement of flank wear of circuit board drills, *Journal of Electronic Packaging*, **114**(1), 93–96.

Rhyne, V.T. (1973) *Fundamentals of Digital Systems Design*, Prentice-Hall, Englewood Cliffs, NJ.

Richards, W. and Parks, E.A. (1971) Model for color conversion, *Journal of the Optical Society of America*, **61**, 971–976.

Ritter, H.J., Martinez, T.M. and Schulten, K.J. (1989) Topology-conserving maps for learning visuo-motor coordination, *Neural Networks*, **2**(3), 159–168.

Roberts, L.G. (1965) Machine perception of three-dimensional solids, in *Optical and Electro-Optical Information Processing* (eds J.T. Tippet *et al.*), MIT Press, Cambridge, MA.

Rodeick, R.W. (1973) *The Vertebrate Retina, Principles of Structure and Function*, Freeman Publishing, San Francisco, CA.

Rosandich, R.G. (1995) Implementation of competitive learning in the HAVNET neural network, in *Intelligent Engineering Systems Through Artificial Neural Networks*, Vol. 5 (eds C.H. Dagli, M. Akay, C.L.P. Chen *et al.*), ASME Press, New York, pp. 173–178.

Rosenblatt, F. (1958) The perceptron: A probalistic model for information storage and organization in the brain, *Psychological Review*, **65**, 386–408.

Rosenblatt, F. (1962) *Principles of Neurodynamics*, Spartan Books, Cambridge, MA.

Rumelhart, D.E., Hinton, G.E. and Williams, R.J. (1986) Learning internal representations by error backpropagation, in *Parallel Distributed Processing*, Vol. 1, MIT Press, Cambridge, MA.

Rushlow, G.O. (1983) Robot operated body inspection system, in *Proceedings of the SME Autofact 5 Conference*, Detroit, MI, Nov 83. Vol. 796, pp. 1–15.

Safabakhsh, R. (1991) Determination of flint wheel orientation for the automated assembly of lighters, *Proceedings of SPIE*, 185–189.

Sarr, D.P. (1991) Aircraft exterior scratch measurement system using machine vision, *Proceedings of SPIE*, 177–184.

Schwartz, E.L. (1980) Computational anatomy and functional architecture of striate cortex: A spatial mapping approach to perceptual coding, *Vision Research*, **20**, 645–669.

Seibert, M. and Waxman, A.M. (1989) Spreading activation layers, visual saccades, and invariant representations for neural pattern recognition systems, *Neural Networks*, **2**(1), 9–27.

Seibert, M. and Waxman, A.M. (1992) Adaptive 3-D object recognition from multiple views, *IEEE Transactions on Pattern Analysis and Machine Intelligence*, **14**(2), 107–124.

Sejnowski, T.J. and Rosenberg, C.R. (1987) Parallel networks that learn to pronounce English text, *Complex Systems*, **1**, 145–168.

Shapiro, L.G., Moriarty, J.D., Haralick, R.M. and Mulgaonkar, P.G. (1984) Matching three-dimensional drawings using a relational paradigm, *Pattern Recognition*, **17**(4), 385–405.

Shepard, R.N. and Metzler, J. (1971) Mental rotation of three-dimensional objects, *Science*, **171**, 701–703.

Silven, O., Virtanen, I., Westman, T. *et al.* (1989) A design data based visual inspection system for printed wiring, in *Advances in Machine Vision* (ed. J. Sanz), Springer-Verlag, New York, pp. 192–212.

Simpson, P.K. (1990) *Artificial Neural Systems: Foundations, Paradigms, Applications, and Implementations*, Pergamon Press, New York.

Skaggs, F.L. (1983) Vision-based parts measurement system, in *Proceedings of the Third International Conference on Robot Vision and Sensory Controls*, Cambridge, MA. pp. 579–589.

Smith, B. (1993) Making war on defects: Six-sigma design. *IEEE Spectrum* **30**(9), 43–47.

Sobh, T.M., Dekhilo, M., Jaynes, C. and Henderson, T.C. (1993) A perception framework for inspection and reverse engineering, in *Proceedings of the IEEE Conference on Computer Vision and Pattern Recognition*, New York, June, 1993. IEEE pp. 609–610.

Stalker, C.J. (1995) A case for PCI in data acquisition. *Instrumentation and Control Systems*, **68**(8), 35–38.

Stilson, D.W. (1956) *A Psychophysical Investigation of Triangular Shape*. PhD Thesis, The University of Illinois.

Sugie, N. and Suwa, M. (1977), A scheme for binocular depth perception suggested by neurophysiological evidence, *Biological Cybernetics*, 26, 1–15.

Sugihara, K. (1986) *Machine Interpretation of Line Drawings*. MIT Press, Cambridge, MA.

Takagi, Y., Hata, S., Beutel, W. and Hibi, S. (1990) Visual inspection machine for solder joints using tiered illumination, in *Proceedings of the Conference on Machine Vision Systems Integration in Industry*, SPIE Vol. 1386, pp. 21–29.

Tarbox, G.H. and Gerhardt, L. (1989) Design and implementation of a hierarchical automated inspection system, in *Proceeding of the Automated Inspection and High-Speed Vision Architectures III Conference*, SPIE Vol. 1197, pp. 13–24.

Tseng, H.F., Ambrose, J.R. and Fattahi, M. (1985) Evolution of the solid-state image sensor. *Journal of Imaging Science*, **29**(1), 1–7.

Uttal, W.R. (1981) *Taxonomy of Visual Processes*. Lawrence Erlbaum Associates.

Van Essen, D.C. (1979) Visual Areas of the Mammalian Cerebral Cortex, *Annual Review of Neuroscience*, **2**, 227–263.

Van Gool, L., Wambacq, P., and Oosterlinck, A. (1991) Intelligent robotic vision systems, in *Intelligent Robotic Systems* (ed. S.G. Tzafestas), Dekker, New York, pp. 457–507.

Vernon, M.D. (1952) *A Further Study of Visual Perception*. Cambridge University Press.

Wasserman, P.D. (1989) *Neural Computing: Theory and Practice*. Van Nostrand Reinhold, New York.

Welch, T.A. (1984) A technique for high performance data compression. *IEEE Computer*, **17**(6), 8–19.

Werbos, P. (1974) *Beyond Regression: New Tools for Prediction and Analysis in the Behavioral Sciences*, PhD Thesis, Harvard University, Cambridge, MA.

Wetherill, G.B. (1969) *Sampling Inspection and Quality Control*, Methuen, London.

White, K.W., Morgan, R.H., McConnell, B.C. *et al.* (1990) The application of high-speed package inspection at R.J. Reynolds Tobacco Company, in *Proceedings of SME Vision '90*, Detroit, MI, Nov 1990. Vol. 13, pp. 1–15.

Widrow, B. (1961) The speed of adaptation in adaptive control systems, in *Proceedings of the American Rocket Society Guidance Control and Navigation Conference*.

Widrow, B. and Angell, J.B. (1962) Reliable, trainable networks for computing and control, *Aerospace Engineering*, **21**, 78–123.

Widrow, B. and Hoff, M.E. (1960) Adaptive Switching Circuits, *IRE WESCON Convention Record*, Part 4 (1960): 96–104.

Winkler, G. (1983) Automated visual inspection, in *Proceedings of the Industrial Applications of Image Analysis Workshop*, Antwerp, pp. 13–30.

Yamamoto, S. (1992) Development of inspection robot for nuclear power plant, in *Proceedings of the IEEE Conference on Robotics and Automation*, Nice, France, May 1992, IEEE, pp. 1559–1566.

Ye, S. *et al.* (1992) Inspecting parts of car body by visual gauging system, *Proceedings of the 38th International Instrumentation Symposium*, ISA, pp. 561–565.

Yoda, H., Ohuchi, Y., Taniguchi, Y. and Ejiri, M. (1988) An automatic wafer inspection using pipelined image processing techniques. *IEEE Transactions on Pattern Analysis and Machine Intelligence*, **10**(1), 4–16.

Zadeh, L. (1965) Fuzzy sets, *Information and Control*, **8**, 338–353.

Zeki, S.M. (1978) Uniformity and diversity of structure and function in rhesus monkey prestriate visual cortex, *Journal of Physiology*, **277**, 273–290.

Zuech, N. (1988) *Applying Machine Vision*, Wiley, New York.

Zuech, N. and Miller, R.K. (1989) *Machine Vision*, Van Nostrand Reinhold, New York.

Index

MIX
Papier aus verantwortungsvollen Quellen
Paper from responsible sources
FSC® C105338

If you have any concerns about our products,
you can contact us on
ProductSafety@springernature.com

In case Publisher is established outside the EU,
the EU authorized representative is:
Springer Nature Customer Service Center GmbH
Europaplatz 3, 69115 Heidelberg, Germany

Printed by Libri Plureos GmbH
in Hamburg, Germany